DIARY OF A CITIZEN SCIENTIST

ALSO BY SHARMAN APT RUSSELL

Standing in the Light: My Life as a Pantheist

Hunger: An Unnatural History

An Obsession with Butterflies: Our Long Love Affair with a Singular Insect

Anatomy of a Rose: Exploring the Secret Life of Flowers

The Last Matriarch

When the Land was Young: Reflections on American Archaeology

The Humpbacked Fluteplayer

Kill the Cowboy: A Battle of Mythology in the New West

Songs of the Fluteplayer: Seasons of Life in the Southwest

diary of a
citizen scientist

chasing tiger beetles
and other new ways of
engaging the world

sharman apt russell

oregon state university press
corvallis

The paper in this book meets the guidelines for permanence and durability of the Committee on Production Guidelines for Book Longevity of the Council on Library Resources and the minimum requirements of the American National Standard for Permanence of Paper for Printed Library Materials Z39.48-1984.

Library of Congress Cataloging-in-Publication Data

Russell, Sharman Apt.
Diary of a citizen scientist : chasing tiger beetles and other new ways of engaging the world / Sharman Apt Russell.
  pages cm
  Includes bibliographical references and index.
  ISBN 978-0-87071-752-9 (pbk. : alk. paper) – ISBN 978-0-87071-753-6 (e-book)
  1. Russell, Sharman Apt. 2. Naturalists–United States–Biography. 3. Women naturalists–United States–Biography. 4. Entomologists–United States–Biography. 5. Women entomologists–United States–Biography. 6. Women college teachers–United States–Biography. 7. Tiger beetles. I. Title.
  QH31.R827R87 2014
  508.092–dc23
  [B]
                           2014023892

Oregon State University Press
121 The Valley Library
Corvallis OR 97331-4501
541-737-3166 • fax 541-737-3170
www.osupress.oregonstate.edu

Contents

Preface

A few notes concerning this book. Why the United States never went on the metric system remains grist for a blockbuster, conspiracy-theory novel–and then the movie starring Tom Hanks. But because we didn't, and because most American readers still see in terms of inches and feet and miles, I use those terms for general observations, switching to millimeters only for the scientific measurement of small things like tiger beetle eggs and tiger beetle parts. Similarly, I refer to temperatures in Fahrenheit rather than Celsius. I use common names instead of Latin and follow the convention of not capitalizing species except for words that are also proper names or otherwise capitalized–for example, Williston's tiger beetle and Douglas fir and Eastern white thorn acacia. My sources are included in endnotes at the back of the book, along with other texts I liked and found helpful.

Acknowledgments

Who could write a book alone–without family and friends and the kindness of strangers? Not me. I want to first thank my husband Peter Russell for accompanying me on field and car trips around southern New Mexico and for filling in, at times, as beetle collector and Kali-impersonator. My neighbors Carol Fugagli and Sarah Johnson babysat my tiger beetle eggs and larvae when, as a working mom, I couldn't always be there for them. My daughter Maria Russell took videos, tiger-beetled herself, and let me enter into her world of third-graders. My friend Shirley Pevernik was a wonderful companion and cowhand on the Rio Grande. Writers Richard Felger, Ann Hedlund, Stephen Fox, and David Remley read all or parts of the manuscript and offered good advice and encouragement over our semi-regular lunches at a local restaurant. In particular, Ann and Richard were generous with their time. Allison Boyd and Ed Greenberg were the only and best members of the How to Become a Leading World Authority Club.

Photographer Dick is unique and the person I could turn to for pictures of very tiny creatures. I want to also thank my agents Peter and Sandy Riva, and JoAnn Collins, for their constant support. Importantly, Peter Riva helped shape the first chapter of *Diary of a Citizen Scientist*–launching the trajectory of a book. At the Bosque del Apache National Wildlife Refuge, biologist Ashley Inslee was crucial in navigating the permit system to collect beetles on the Rio Grande. Ted MacRae and his blog were a source of inspiration and good information. At Western New Mexico University, where I have taught writing skills for many years, Randy Jennings generously showed me how to preserve my first larval specimen, Manda Clair Jost set me up with a microscope for dissecting beetle ovaries, and Bill Norris provided good cheer and contacts in the bug world online. Of course, the staff at WNMU's Interlibrary Loan are essential and always a pleasure to work with. I'm grateful for Terry Erwin for providing me access to the tiger beetle collection at the Smithsonian Institution; for Jennifer Shirk and Rick Bonney for their energy and time at the first Public Participation in Scientific Research conference in Portland, Oregon; for Jake Weltzin at Nature's Notebook and his support of this project; for Darlene Cavalier and her tireless work in the field of citizen science; for Bob Schiowitz and Elizabeth Toney for squiring me around archaeological sites in the Gila National Forest; for Chris Hass and her help and corrections about coatis; and for Sonnie Susillo for sending me lively e-mails about her tracking experiences. Photographers Elroy Limmer and Dennis Weller provided some extraordinary images for this book. My publisher, Oregon State University Press, has a wonderful staff–thanks especially to Mary Braun, Brendan Hansen, Micki Reaman, Erin New, Tom Booth, Steve Connell, and Maya Polan. Parts of this book were published previously in *Orion Magazine*, *Onearth Magazine*, *High Country News,* and *Writers on the Range.* More thanks go to the editors there: Jennifer Sahn, George Black, Michelle Niejus, and Betsy Marston. Finally, of course, I could not have written this without the help of my entomological mentors, Barry Knisley and David Pearson.

Introduction: Renaissance and Revolution

It's 2007 and you're a young astrophysicist on your third pint in an English pub, clothes rumpled, head in your hands. You whine: in order to prove your latest theory on star formation, you need to compare large samples of galaxies with elliptical and spiral shapes. What you have to work with are a million unclassified galaxy images from a telescope in New Mexico. The shapes of galaxies are patterns computers cannot easily recognize, and you've spent a week, twelve hours a day, sorting through fifty thousand photographs. You close your weary eyes. You can't keep up the pace. A friend murmurs, "Maybe you should get some help?"

Only a year ago, NASA's Stardust@Home project started posting images online from its interstellar dust collector, and citizen scientists eagerly began looking for stardust particles. Could people be trained to classify galaxies, too? You brighten up. A British hurrah. You publicize your idea, you set up the website, and within twenty-four hours, you are getting almost seventy thousand classifications an hour. In the first year, you will get fifty million.

That's the apocryphal story behind Galaxy Zoo, a citizen science program that has since resulted in dozens of peer-reviewed scientific papers, as well as discoveries like "green-pea" galaxies, which produce stars at a high rate and may help us understand how the first stars formed. Each galaxy classification by a single volunteer is corroborated by thirty more volunteers, with a resulting accuracy equal to that of professional astronomers. The universe—which may contain as many as five hundred billion galaxies—is slowly being mapped by cartographers of all ages, all occupations, and all nationalities.

Around the world, citizen science projects are proliferating like the neural net in a prenatal brain. The sheer number of citizen scientists, combined with new technology, is beginning to shape how research gets done. Some 860,000 people have participated in Galaxy Zoo and related projects on the website Zooniverse. More than a quarter million play the video game Foldit, helping biochemists synthesize new proteins. The

National Geographic Society's search for archaeological sites in Mongolia sends satellite images from the field to thousands of citizen scientists downloading them at home. The use of crowdsourcing to take advantage of large numbers of human eyes and brains has inspired the development of algorithms to improve how computers themselves work; like Yoda, we can teach them our mysterious ways.

Although the biggest citizen science programs are online, many other citizen scientists are getting up from the computer, going outside, and joining a research team to study urban squirrels or phytoplankton or monarch butterflies. Most obviously, they help scientists count things: juniper pollen, comets, horseshoe crabs, dragonfly swarms, microbes (in your belly button and in your kitchen), picas, thunderstorms, roadkill. An estimated two hundred thousand people work with the Cornell Lab of Ornithology tracking and monitoring birds, with over a million observations reported each month on the Lab's online checklist. Citizen scientists also double as environmental activists, collecting air and water samples, documenting invasive species, and looking at changes in species behavior.

An army of human volunteers has become an army of scientific instruments, and that's not a new idea. In China, people have been recording locust outbreaks for over three thousand years. French wine growers began tracking grape harvests in the fifteenth century. Charles Darwin relied on a network of amateurs for observations of the natural world, working-class men and middle-class women, vicars and shopkeepers with whom he corresponded by penny post. Today we've replaced the pen with the login, using the Internet to communicate in ways that make large-scale, long-term projects possible.

One of the newest and potentially most important branches of citizen science is the analysis and understanding of global warming, with programs like Nature's Notebook and Project Budburst using volunteers to monitor plant and animal responses to a changing climate. What plants are budding when? What birds are here now? What insects have emerged?

In Portland, Oregon, a couple and their two children walk the trails of urban parks watching for the first leaves of kinnikinnick, the flowers of Indian plum, the fruits of the mellifluous salmonberry, snowberry, thimbleberry. Before the start of spring, the trilliums are underground and the snowberry leafless. Suddenly the Oregon grape has clusters of yellow flowers that attract hummingbirds. Warmer and longer days bring more color and scent, camas lily and bleeding heart and lupine and salal and wild rose, and then in the fall, maples turn red and the leaves of Solomon's seal yellow and gold. The oldest daughter records this information online. Carefully, the family marks the appearance of spotted towhees and northern flying squirrels, the absence of Pacific tree frogs. Their efforts are being duplicated across rural and urban America by thousands of men, women, and children.

This is renaissance, your dentist now an authority on butterflies and you (in retrospect this happened so pleasantly, watching clouds one afternoon) connected by Twitter to the National Weather Service. This is revolution, breaking down the barriers between expert and amateur, with new collaborations across class and education. Pygmy hunters and gatherers use smartphones to document deforestation in the Congo Basin. High school students identify fossils in soils from ancient seas in upstate New York. Do-it-yourself biologists make centrifuges at home. This is falling in love with the world, and this is science, and at the risk of sounding too much the idealist, I have come to believe they are the same thing.

My own work with tiger beetles, under the guidance of two generous mentors, was done mainly during the field seasons of 2011–2012. The entries that make up this book describe that fieldwork and have been shaped from written notes and the observations of those two years. In the larger world of citizen science, not much has changed from then to the writing of this introduction now. Only the numbers have increased: more and more people are watching birds, taking water samples, staring into the heart of a red spiral galaxy, marrying curiosity with collective power, waking up and thinking–what am I going to study today?

Bronzed tiger beetle (photo by Ted MacRae)

July 2011

July 23

I'm fifty-seven years old today, squatting on a sandy riverbank, watching a pack of Western red-bellied tiger beetles eat a dead frog. Although the insects are two feet away and only about a third of an inch long, through close-focusing binoculars they fill my vision and I see an entirely new and surprising world. Tiger beetles have disproportionately large, sickle-shaped mouthparts, which they use to stab the white belly of the frog, slicing and scything and scissoring their mandibles like a chef sharpening his knives. Sometimes the beetles stand completely still, as if to pose, each brown wing cover patterned with seven creamy irregular dots, the abdomen orange, the head and thorax iridescent in the sun. The beetles flash red and green and blue and gold. Suddenly they are gone. Suddenly they return. Suddenly they stare straight at me, their large bulging eyes giving them an inquisitorial air. Perhaps thirty of them feed on the frog's slightly bloated carcass, and I am reminded of lions at a kill, although lions don't look half so fierce.

I have always wanted to be a field biologist. I imagine Zen-like moments watching a leaf, hours and days that pass like a dream, sun-kissed, plant-besotted. I imagine, like so many others before me, a kind of rapture in nature and loss of ego. John Burroughs, an early American naturalist, wrote that he went to the woods "to be soothed and healed, and to have my senses put in tune." In my own walks through the rural West, this echoes my experience exactly. I enlarge in nature. I calm down. The beauty of the world is a tangible solace—that such harmony exists, such elegance, the changing colors of sky, the lift and roll of land, a riverbank, and now a beetle flashing in the sun, an entrance into its perfect world. I am soothed, I am thrilled, and at the same time, eventually I get bored. Eventually I go home because my work (my writing, my students, my laundry) is elsewhere.

But what if that employment, my engagement with the world, was right there, in the largeness and calm of nature itself? "Blessed is the man,"

Burroughs continued, "who has some congenial occupation in which he can put his whole heart, and which affords a complete outlet to all the forces there are in him."

I have always wanted to be John Burroughs, and I have always wanted to be Jane Goodall, who left her home in England–not even going to college first–to work as a secretary for anthropologist Louis Leakey and later to study chimpanzees in Tanzania. She lived in the forests of Gombe, her back against a tree, her toes rotting with fungus, beset by mosquitoes, watching and listening, entering the world of forest and animal–and always, always taking notes, desiring and finding and then opening "a window" into the mind of another species, one not "misted over by the breath of our finite humanity."

Sometimes in the middle of the street, in the middle of my life as a teacher and writer and wife and mother in southwestern New Mexico, I have stopped to wonder: Why didn't I do that? Why didn't I go to Africa? It's a sorrow. My heart actually feels pierced. Where is my window into the unknown, the nonhuman? And where is my competence? My expertise? My forest? Why am I inside so much of the day?

Through close-focusing binoculars, I watch a female Western red-bellied tiger beetle eat a frog. I know she is female because of the slightly smaller male riding her back. Likely, this male is trying to be the last to introduce his sperm into her sperm storage compartment; instead of eating, instead of watching for things that might eat him, he holds on. A second male tiger beetle approaches the pair, perhaps to try to disengage the first, and then from somewhere the music rises and reaches a crescendo as a giant, warty, oval, gravel-colored toad bug–known for grabbing its prey with sturdy forelegs and sucking out their vital juices–attacks the second beetle, jumping on him and as quickly jumping off, for the toad bug and the tiger beetle are about the same size, and they are both predators with the mouthparts to prove it. The toad bug lumbers away, and the tiger beetle shakes itself as if relieved, although really it is only stilting, lifting up on its long legs to cool off, getting a fraction farther away from the hot ground.

I know about stilting, just as I know that the fastest insect in the world is an Australian tiger beetle, who can gallop 5.5 miles an hour or 170 body lengths per second, because I have recently read *Tiger Beetles: The Evolution, Ecology, and Diversity of the Cicindelids,* co-authored by

David Pearson, a conservation biologist at Arizona State University, world expert on tiger beetles, and advocate for the citizen scientist–the unpaid, not professionally trained, non-scientist. Particularly in the field of natural history, more and more people armed with better and better field guides and smartphone apps are now able to gather information on a species' distribution, behavior, and biology. Fewer scientists need to duplicate that work. As David Pearson has told me without chagrin, many areas of traditional research in biology have been "largely turned over to amateurs."

"They are the future," David says. "People like you."

David and I are in regular contact because when I e-mailed him a question about citizen science, he e-mailed right back. David believes that conservation biology is in danger of losing its relevance–its ability to affect policy and protect diversity–if scientists do not reach out to the general public, abandoning their exclusive jargon and welcoming the barbarian hordes. A tall, genial man in his sixties, he has been effortlessly kind to me, sending papers and links to websites and urging me toward a study of tiger beetles, which he points out are bio-indicators of biological diversity, since where tiger beetles thrive, other species of birds and butterflies tend to thrive, too. In his own professional work, David has used a census of tiger beetles to help set the boundaries of a new national park in Madagascar and assess the status of a protected area in Peru. Prodding other people into spending money and time on his favorite insect is something David does on a regular basis: the lawyer in Cambridge now writing a book on the tiger beetles of Mexico, the dentist in Ohio with his fabulous collection of North American tigers, the new guy in Texas who has become somewhat obsessed.

I'm an easy mark. Ten years ago, I was first inspired by another entomologist, Dick Vane-Wright, the Keeper of Entomology at the London Museum of Natural History, whom I was interviewing about butterflies. "There's so much we don't know!" Dick told me, sounding excited and distressed at the same time. "You could spend a week studying some obscure insect and you would then know more than anyone else on the planet. Our ignorance is profound."

Nodding, I wrote the comment down in my notebook. I liked its humility–an acceptance of how little we know–and I liked its challenge and implied sense of wonder–there is still so much to discover. Over the next decade, the words would surface again, like some message on a Magic Eight Ball: *Signs point to yes. Concentrate and ask again. You could spend a week studying some obscure insect and you would then know more than anyone else on the planet.*

I've spent a lot of time in my life, much more than a week, thinking about how terrible things are on the planet: how polluted, how crowded, how damaged and diminished. In my circle of friends, the apocalypse is party conversation. Dead zones in the ocean. The melting ice caps. Then there are the changing patterns in our own weather–that unusually dry winter followed by a dry spring. Global warming is local, with most of the Southwest in what is called exceptional drought condition, the highest category of drought, one expected to persist and intensify.

I've lived in the desert almost all my life and waiting for rain is nothing new. Every summer in the Gila Valley of New Mexico, we watch the skies, anticipating the monsoon season of July, August, and September, which provides us with half our annual precipitation of twelve inches. But here it is, mid-July, and the rainfall this year–counting all the way back to January–amounts to less than my little finger, a brittle twig, a shortened stalk of grass. This year has also been a bad fire season, with over a dozen homes destroyed in the town of Silver City, where I teach, and hundreds of thousands of acres burned in the surrounding Gila National Forest and nearby Arizona. My friends and I are fearful of change and perversely excited by its drama at the same time.

As the world falls apart, as we lose hundreds of species a day in the most current mass extinction, as I lift my head to the bright blue New Mexican sky and lament and wail and ululate … the idea that there is still so much to discover strikes me as a kind of miracle. We think we've beaten the Earth flat, hammered out the creases, starched the collar, hung her up to dry. We've turned the planet into our private estate, a garden here, a junkyard there, maybe an apocalypse at the end. But no longer wild, no longer mysterious. And yet. *You could spend a week studying some obscure insect and you would then know more than anyone else on the planet.* It's such a cheerful thought.

Earlier this month, as things on the planet kept getting steadily worse, David Pearson e-mailed me that the salt lakes in northeastern New Mexico had some interesting tiger beetle species and that "studying *Cicindela*

Pile-up of Western red-bellied tiger beetles (photo © Joseph Warfel/Eighth-Eye Photography)

sperata at the junction of Highway 60 and the Rio Grande River would be especially intriguing."

Much like humans, tiger beetles can be found almost everywhere, except for a few spots such as Antarctica, the Arctic north of 65 degrees latitude, and isolated islands like Hawaii and the Maldives. Some twenty-six hundred species of tiger beetles have been described so far. The Rio Grande tiger beetle or *Cicindela sperata* is a species common in the American Southwest, not endangered and not threatened. Still, we don't know much about this insect. Still, this beetle is a mystery, a blank spot on the map of tiger beetles in the world.

All tiger beetles share some characteristics. Tiger beetle larvae are pale white grubs with a dark, armored head; heavy, knifelike mandibles; and up to six eyes capable of close focusing. After hatching from very tiny eggs (2-4 millimeters) typically laid into soil, the very tiny larvae use their tiny jaws to dig a narrow tunnel, the head acting as a shovel. On the larvae's lower back, two pairs of hooks anchor them into the side of that tunnel—allowing the grubs to lunge out like B-movie monsters and drag in insects even smaller than themselves. Inside the tunnel, the prey is dismembered and eaten, and the indigestible parts carried up and thrown from the burrow's mouth. Over months or a year or two years or four years–going into dormancy in the winter–these larvae go through three stages, or instars, during which they eat and grow larger and shed their skins, each time digging the tunnel wider and deeper to accommodate their new size. Finally, the last stage forms the pupae from which the beetles emerge.

But in what kind of soil or habitat does the Rio Grande tiger beetle lay its eggs? After the eggs hatch, what do the larvae of this species look like exactly? And how long is the cycle from egg to larva to pupa to adult? David Pearson was pushing me to answer some of these questions. The co-author with Barry Knisley of *A Field Guide to the Tiger Beetles of the United States and Canada,* he wanted to replace the slightly embarrassing phrase "larval biology unknown" that marred his entry for the Rio Grande tiger beetle.

Caught up in the e-mail moment, flattered to be considered a potential citizen scientist, I thought, yes, that *would* be intriguing. That would be fun.

Which is why I am here today, having decided to go outside and actually see a tiger beetle–live, not in a book. Tiger beetles in my area are not active as adults until after the summer rains. The Western red-bellied tiger beetle is an exception, appearing before the monsoons and congregating in groups along riverbanks, ponds and irrigation ditches, stock tanks and mud flats. The beetles I am now watching stab a frog are behaving in full accord with *A Field Guide to the Tiger Beetles of the United States and Canada*, which notes, "As water sources dry up and tadpoles, insect pupae, and other aquatic organisms become exposed, hundreds of Western red-bellied tiger beetles will attack these large but helpless organisms."

Attacking a large, helpless organism is high on any tiger beetle's list. But more often these insects are hunters, not scavengers, the adults running down an ant or spider, grabbing the victim, and drenching its body with digestive juices. The Western red-bellied beetle can also fly short distances, and I am reminded that this bank of the Gila River, about two miles from my home, contains many more tiger beetles than the small group I am watching. The area swarms with carnivorous beetles, just as the guidebook promised. Putting down my binoculars, I can see them on the mud at the water's edge. I can see them in the grass and in the sedges. I can see them everywhere in my peripheral vision, and I think of piranhas and certain horror movies I watched as a child. It's silly to be at all nervous. Even so, I look quickly at my feet.

I am fifty-seven years old today and have lived much of my life during the last half of the twentieth century, surrounded by stories from movies and TV and books, of course, so many books, making my decisions (not going to Africa), reading and writing and living mostly in my head, the movie of Sharman. In this way, the years flew by. And here I am now watching tiger beetles on a sandy bank of the Gila River, in good physical shape, with a lively mind and two grown children and a few books I have written myself. Many decisions are no longer mine to make. I no longer hope to become an ER doctor or the creator of cool television shows. I don't plan on a career in the United Nations or as a white-water rafting guide. At every point in life, and not just the tail end of middle age, there is a long list of what we can no longer be: we are defined by our limits as much as our loves.

This speaks to the power of transformation. Turn the idea around. At every point in life, there is a long list of what we can still be. I can still choose a window into the unknown, the thrillingly nonhuman. In every moment of the day, in the middle of any day, I can become newly engaged with the world. Newly competent. There's so much to discover! So much we don't know. I can still become something I am not.

Can't I?

July 28

I feel accomplished just contemplating the mound of equipment in my home office, where I write and teach online classes about writing. The four plastic terrariums the size of my toaster oven and the six terrariums half that size are for the Rio Grande tiger beetles I plan to catch and hold captive, watching them mate and lay eggs, and then watching the eggs hatch and grow into larvae. The two collector nets have expandable handles that extend to twelve feet. The assorted clip-on desk lamps are for heating the terrariums to mimic the sun and temperature of a summer's day on the riverbank. I also have extra boxes of lightbulbs in varying degrees of brightness. Temperature monitors. A bag of weird blue polymer gel called insect water. Lengths of clear plastic tubing to be cut into individual larval rearing tubes, which will also require wire mesh and duct tape. A wind-up, no-batteries-needed fluorescent lantern for attracting insects at night. A GPS I don't know how to use. My close-focusing binoculars, which I don't know how I ever lived without. Hat. Nicer hat. Water bottle. Nicer water bottle. Backpack. Trowel. Insecticide. Flashlight. Feather-weight forceps. It's been a consumer frenzy. Like tiger beetles on a dead frog.

This morning I e-mailed David Pearson to say that as well as checking out the bridge on Highway 60, I have permission to collect the Rio Grande tiger beetle at the Bosque del Apache National Wildlife Refuge thirty miles south of that bridge. This is an overnight trip, a four-hour drive from my home in the Gila Valley. I'm not in a hurry to rush over. The Rio Grande tiger beetle is active after and during the summer rains, and I'm waiting for the monsoons to fully start, anticipating that perfect moment when the beetles are abundant and easy to catch, even for someone who swings a collector's net like she once played softball in the fifth grade.

But David has no patience for waiting and shoots back this reply:

From: David Pearson
Sent: Fri, 28 Jul 2011 12:19:13 PM
Subject: RE: quick question
If there has been even a little rain, the tiger beetles along the
Rio Grande have probably emerged and will only stay active for
about six weeks. The only guarantee is if you don't go over there,
you won't see any. Good luck. Dave

July 29

The biologist in charge of my collector's permit at the national wildlife refuge has left me a can of bear spray. The refuge has had some "problem bears" in the area, as well as "problem mountain lions," and that's where the biologist is at this moment—trying to radio-collar a mountain lion. The volunteer at the visitor center explains how to use the spray, although I can't imagine having the wherewithal to fumble in my backpack (I've had wrestling matches with my purse that border on domestic violence), get out the can, unlatch the lever, and aim for the eyes. But I listen politely. Cock here. Pull back. Later you can use the bear as a flotation device.

I'd love to see a bear today. I'd love to see a mountain lion, an animal that may remind us of a domestic cat but on such a large-scale, able to run forty miles an hour, drop silently from a height of sixty feet, land running, then leap fifteen feet and spring three times that. Mountain lions were hunted as pests and cattle-killers through much of the twentieth century, until by the late 1960s they had been eliminated in the eastern United States and were declining elsewhere. Then some states began to reclassify them as big-game animals eligible for trophy hunting. Attitudes about nature had also begun to shift in an increasingly urbanized society–and mountain lion populations rebounded. Although these solitary animals are notoriously hard to count, we can guess we now have about twenty to thirty thousand in the continental United States, with the numbers slowly expanding.

In a familiar conversation in the rural West, I've discussed getting eaten by a mountain lion with a number of friends. One was a writer for *National Geographic* and so maybe he had to say he would not enjoy, but certainly appreciate, the experience as a kind of achievement. There are worse and more banal ways to die. I have to agree–even while imagining the heart-stopping, heart-racing exigency of the body. Screaming. Urination. Blood. Pain. A kind of achievement.

Last spring another friend of mine, camping in the Gila Wilderness, was almost eaten by a mountain lion as he rested in the grama grass and watched squirrels weave bushy tails through the oak branches above. Hearing a noise, he rose up on one elbow. The lion was ten feet away. My friend was immediately up and walking backwards with the animal follow-ing. Trotting. At this important moment, my friend resisted the urge to run and instead stopped, yelled, and growled. His wife, who was nearby in camp, said she heard a very loud growl. Mountain lions are cautious when it comes to attacking prey. Much like you and me, they don't want to get hurt and evaluate carefully how much trouble this meal might cause them. (I feel the same way about French cooking.) As soon as my friend made himself look big and fierce, a pain in the ass, the mountain lion left. My friend's wife remembers her husband coming back to their tent and chanting, "A fucking mountain lion, a *fucking* mountain lion, a fuck-ing *mountain* lion" although he remembers only saying this once. As in, "Gosh, honey, a fucking mountain lion."

A can of bear spray lodged at the bottom of my pack, I'm off to catch tiger beetles without any real hope of getting eaten by anything except mosquitoes. This actually worries me since I have a visceral reaction to the whine of these insects. The Bosque del Apache Refuge's website has a Mosquito Activity Forecast and last I checked, the hourly mosquito update was moderate, with "conditions that may keep some mosquitoes at bay." I bring along my own conditions, which include mushroom-cloud doses from an aerosol spray backed up by wipes coated in poison.

My husband, Peter, is with me for this trip, and we drive through a locked gate for which I have been given the combination. We park on a levee close to the Rio Grande. From here to the river is a thick stand of salt cedar and cottonwood and willow, indeterminate shrubs, branches, stickers, and insect swarms like holograms dizzily out of focus. We head into this jungle carrying our collector's nets with their long expandable handles.

There is a way to approach a river that resembles birth. A dark tunnellike passage and then emergence into light and water and sun striking water and light filling up the world. The riverbed is wide, a hundred feet across, strips of brown mud and white sand separated by braided streams. The other side of the bank, with its own line of salt cedar and willow, seems far away. The sudden space is startling.

In 1598, Gaspar Pérez de Villagrá described the Rio Grande or what was then called "The River of the North" while crossing it with the Juan de Onate expedition on their way to claim everything for Spain. Villagrá, who had been educated at the University of Salamanca in Spain, was just then composing the first written epic poem of the United States, *Historia de la Nueva Mexico* (1610), with thirty-four cantos chronicling the hardships, successes, and cruelties of the expedition. In one of those, he marveled at the river's "peaceful, suave, pleasing and mild water" spread out over a wide flat area, with many kinds of fish and wild birds such as cranes, ducks, and geese. The soldiers, who had thought themselves lost, now felt they had entered a kind of Elysian Fields, the final resting place in Greek mythology for the souls of the heroic. At first, they overindulged, and "consumed by the burning thirst, their tongues swollen and their throats parched, threw themselves into the water and drank as though the entire river did not carry enough to quench their terrible thirst." Afterwards, Villagrá reports disapprovingly that they lay on the cool sands "like foul wretches stretched upon some tavern floor in a drunken orgy, deformed and swollen and more like toads than men." Two of the horses also drank so much that they died, and two others plunged into the stream, were caught up in the swift current, and drowned.

More than four hundred years later, the Rio Grande has been dammed and parceled out like stacks of money after a bank heist, and what I see today is what remains after the irrigators upriver get their share. Even so, I step from the bank almost immediately into a channel that runs fast and brown, ankle- and then thigh-deep. Whoa! Wet! Keep the GPS dry! I am stumbling, almost falling, before lurching upward onto white sand. Sun and light and my shoes, socks, legs, and shorts are now filmed with a thin coating of mud, the muddy water of the Rio Grande where I nearly got to wash my face and hair too.

I must stop for a moment—just to collect myself. Then I move forward slowly, scouring the ground for tiger beetles or for the movement of tiger beetles because that is how you first understand them, as a blur of flight, a disturbance in the air ahead or to the side. Your eyes refocus and then you may actually see the insect, a tiny insignificant dot on Earth—a dot uniquely patterned in a curious raised pose, stilting up on its long, long legs.

My husband follows behind, swishing his net back and forth. "When should I do the obligatory try-to-catch-you-with-this?" he asks.

"Never," I say, because the truth is that I'm not very good at physical play. It's hard for me to giggle and tickle and chase and have that kind of spontaneous fun. I was too serious as a child, and I can be too serious now, living in my head and the sorrows therein.

But I feel a joy here. I feel that brightness in the veins, in the chest. I have a purpose here, surrounded by beauty, by water, by light. I put down my pack with its bear spray and collecting boxes and water and sandwich, and I feel light and easy, and I swing my collector's net just a little bit, like a flag waving.

✳ ✳ ✳

David Pearson has warned me. Catching tiger beetles requires the stalking movements of a great blue heron or Tai Chi master. Sometimes it is necessary to crouch, inching forward so as to not frighten the beetle, angling the body so as to not let one's shadow signal one's approach. Think pure thoughts. Think flow. And be prepared to slam the net down before the high-energy, Type-A tiger beetle suddenly flies or runs off in pursuit of prey or water or some private exultation. David is torn between two techniques, the mighty over-the-head swing or the twist of the wrist that flips the net sideways over the insect. The first is harder to control and can produce a telltale shadow; the second requires a closer approach.

The problem I have is shifting gears from slow and flow to jumpy and hoppy, slamming the net down and rushing over in a half-crouch to make sure all parts of the net are pressed firmly into sand or mud. The one-third-inch-long tiger beetle will escape through any gap. This doesn't usually happen since most often the net swings down and there is no insect beneath it–the tiger beetle has already escaped while the net was in the air. I swing ten times for every beetle I trap, and half of those then slip under and away.

For those I do catch, I lift the top of the net, forming a cone. The beetle flies up and I corner it further in a small fold of cloth, pinching the area tight with my fingers. I'm prissy, careful to avoid getting bitten by those mandibles, not keen on handling the insect or feeling it against bare skin. Above all, I don't want the sound of small parts breaking–to see the beetle's insides or the awkward angle of a wing. I am not being nice, just squeamish.

From the net I transfer the beetle into a collecting box, pushing the fold of cloth and its captured insect into the box and then withdrawing the net, leaving the beetle behind and quickly shutting the lid. A few more insects escape during this process.

Several hours will pass before Peter and I have ten tiger beetles to take home. This would not be possible if there were not so many tiger beetles up and down these sand and mud banks, posing, stilting, darting, flying away. They are not, unfortunately, the Rio Grande tiger beetle I have come to study–at least, I don't think so. Yes, the pattern on their wing covers is similar to the pattern of the Rio Grande tiger beetle: two cream-colored C shapes facing each other on the top third of the wing cover, two loops like musical notes in the middle third of the wing cover, and two fillips on the bottom third of the wing cover. But according to the guidebook, the Rio Grande tiger beetle has more of a J than a C at the top. More obviously, the underside of these beetles is flashing metallic blue, while the underside of the Rio Grande tiger beetle is a dark copper color.

Squinting at the photos in the field guide, I tentatively identify this species as the bronzed tiger beetle, which an expert on tiger beetles might have immediately distinguished from the Rio Grande tiger beetle by its bulkier body and shorter legs. I also see what I believe are ocellated tiger beetles with a simple pattern of four cream-colored dots on brown.

Everything I see feels doubtful. Is that an inverted J or a C–or even a G? Is that metallic blue or the sun glinting off a copper underside? What is "shorter," exactly, and "bulkier"?

Could this, instead, be the coppery tiger beetle or the coppery form of the aridland tiger beetle–sometimes colored a bright green?

I am taking my insects home to identify them and also to see if I can keep them alive. I'm reading the guidebook, at the center of the world, the center of sun and water and light, when suddenly I jump in my skin. My body already knows the sound of riffle and birdcall. But this is different and close, and I turn–shifting gears from reading to fleeing–to see a large fish struggling upstream in a shallow braid of water. Brown water slaps against the raised back, and I assume this is the common carp, which can reach up to thirty pounds. Historically, shovelnose sturgeon and longnose gar and American eel also swam in the Rio Grande, but these species have long been replaced by exotics like this one.

Two more times today, I'll be startled by this wet slap-slap-slapping. Each time, I hope for a bear.

✳ ✳ ✳

With their patterned wing covers and jewellike colors, tiger beetles have a cachet among entomologists and insect collectors. David Pearson confirms there is a "tiger beetle mania" which seems to strike "otherwise quite normal" people. I suspect this charisma is connected to the beetle's aggressive hunting skills. Their good vision–those prominent eyes–allows them to spot and pinpoint the movement of smaller insects. Their long legs, inserted into pivots along their body, give them speed–and so they are off, running in short fast spurts, running so fast that soon they can no longer see what they are chasing. A brief stop allows their eyes to refocus on the prey's movements. Another spurt and the attack is messy, the remains raked into the tiger beetle's mouth cavity and masticated by rear teeth into a puree that the beetle sucks up through a straw-like organ, spitting out the indigestible parts.

Like solar batteries, most tiger beetles are revved up by the sun, using outside temperature to control their inner temperature, which they like to keep at just below the lethal limit of 116-120 degrees Fahrenheit. This allows maximum speed for hunting prey, seeking mates, and escaping predators. Except for nocturnal species, the best time to find tiger beetles is on a sunny day with temperatures above 77 degrees Fahrenheit.

If Peter and I leave the wildlife refuge now, we have time and sun to head north to look at more sites along the river. Running parallel to Interstate 25, the Rio Grande is accessible from frontage roads and the bridges over state highways 60 and 44, and we take these roads, which I've never had reason to drive before, only catching glimpses from the freeway. The view close-up is not appealing: a string of homes unconnected to any town or source of income, mostly trailers, hard to heat and hard to cool, depreciating in value every second of the day. One out of five New Mexicans lives below the poverty line, and I know what that can look like. It looks like tires, walls of tires in your front yard. It looks like cars that don't run and refrigerators that don't work. It looks like stuff: used, broken, beyond hopeful.

Driving almost next to the river now, we also see alfalfa fields and a few big farmhouses, banks of sunflowers, wealthy people with horses and not so wealthy people with horses, paddocks of bare, beaten earth. We go to David Pearson's bridge intersecting Highway 60. We go to Escondido Lake Park, a good-sized pond next to the river, and to two other promising river edges. Everywhere we go, we don't see tiger beetles. And that's surprising.

In 2001, Barry Knisley (co-author of *A Field Guide to Tiger Beetles of the United States and Canada),* wrote a report for the US Fish and Wildlife Service on tiger beetles on the middle Rio Grande. He looked at historical records and conducted surveys at twenty-five sites, pinpointing the Bosque del Apache National Wildlife Refuge as the one with the greatest diversity and abundance—eight species and thirteen hundred individual beetles noted by two workers walking each for sixty minutes along the sand and mud bars of the river. I've been e-mailing Barry for advice and have just visited four of the sites in his report, places where he easily saw the Rio Grande tiger beetle, *Cicindela sperata.* In retrospect, that 2001 field season must have been the Rose Parade of Tiger Beetles: the river flush, air in motion, undersides flashing, musical notes, and mandibles. The winners of that count: 1,675 Western red-bellied tiger beetles, 1,579 bronzed tiger beetles, 591 thin-lined tiger beetles, and 355 Rio Grande tiger beetles. Today, ten years later, only the Bosque del Apache has any beetles at all.

Tiger-beetling is a great excuse to poke around in the world. Escondido Lake Park would be a sad place to visit—no money for maintenance in these tough days, trash along the pond, signs with peeling paint–except for the man in a ball cap teaching his granddaughter how to fish. Except for the family with a Saturday picnic, radio playing, kids playing. Tiger beetles like sun and so do people, and we are all out and about, darting here and there in our private exultations. The horses bend their long faces masked against flies. The sunflowers are assertively cheerful. And driving past one more junkyard, I think optimistically that all these tires *will* be useful someday, recycled and cannibalized, serving meanwhile as a good home for mice. For the moment, serving someone's psyche, proof of ownership where ownership says it all: at least I have my 1,098 car parts.

At least I have my two boxes of ten beetles, and we stop to give them live prey to eat, mealworms that I've carried in a leaky cooler. My hands fumble with the container so that the lid pops and dozens of grubs spill out into the crevices of our Honda Civic. Immediately I try to forget this has happened. By tomorrow it will not have happened at all. (Luckily I rarely sit in the back seat.) For water, I use the weird blue chunks of polymer gel, the modern version of an insect drinking bowl. My shoes and clothes are heavy with mud. I can lick my lips and taste insecticide. I can close my eyes and remember that emergence into water and light, sun on water, and the song, of course, "I'm an Old Cow Hand (from the Rio Grande)," the song I've been singing all day.

July 30

The next morning we drive east to Laguna del Perro, one of the largest salt lakes dotting the desert of eastern New Mexico. We pass more trailers–each one this time on twenty or more acres, each surrounded by a fan of still-good-stuff drying out in the sun and used-to-be-good stuff ruined by rain–part of a housing development for the poor and romantic on this flat plain sixty miles from the city of Albuquerque. When gas was cheap, maybe they commuted. Maybe they are retired or live on disability. Maybe they cook meth or did; we go by a number of burned-out trailers. Something drew them to this empty landscape stretched and tacked to the horizon. On a giant chessboard, the pieces in place, each trailer represents a separate drama–the dream of country living, a passion for space, misanthropy, a cheap place to live. I'm trying to imagine their neighborhood potlucks.

David Pearson has urged me to visit Laguna del Perro because the larval biology of the tiger beetles here is relatively unknown and what he calls "extreme." I've read about his study of one species, the Williston's tiger beetle, whose larvae (at least of three subspecies) build over their burrows chimney-like towers with two projections at the top. These medieval turrets are ten to twenty millimeters high and clearly visible on a level lake bed or mud flat. In some cases, they must look like the trailers we have just passed, carefully spaced apart, each home a castle. We all find a way to live on this planet.

David and his colleague Barry Knisley were intrigued. They hypothesized that the turret helped the larvae survive flooding during the monsoon season; the walls stopped water from drowning the grubs in their burrow and the larvae could still wait for prey and hunt at the top of the tower. To test their theory, the scientists artificially flooded one hundred turrets. Every turret collapsed. They used a watering can to mimic a hard rain. Again the turrets collapsed. Each time, undeterred, the larvae rebuilt them within a day.

Next the men wondered if the turrets had a thermoregulatory function, cooler at the top in high desert heat. Bingo. A microprobe revealed that from 10:00 a.m. to 4:00 p.m., the soil surface reached above 113 degrees

Laguna del Perro (photo by Peter Russell)

Fahrenheit while at only 20 millimeters above that surface–with less friction and faster wind–the temperature was not over 100 degrees.

David and Barry's third hypothesis was that castle turrets also acted as a defense against enemies. But in testing this idea, they learned that tiger beetle larvae with towers actually had a greater chance of being found and killed by parasitic wasps and flies than the larvae of species without towers. Larvae in towers next to other larvae in towers also seemed to attract more parasites than larvae in isolated towers. The evolutionary choice: cooler temperatures but more attacks by parasites who lay eggs in your body, eggs which hatch and eventually eat you alive; or fewer parasites but a higher chance of desiccation and overheating.

The men would later discover that turrets attracted more prey as well. Prey attraction and temperature regulation seemed to be the key advantages to castle-building.

Irrespective of all that–what fun these guys had! On their hands and knees in the mud, fooling around with watering cans, making up predictions and testable ideas. Build up the tower and admire its elegance. Build up the tower and try to knock it down.

For tiger beetles, the unique habitat of a salt lake can produce distinct populations within a species since groups from one lake bed to the next evolve slightly differently in response to the harsh environment. David and Barry had been studying a subspecies of Williston's tiger beetle in southeastern Arizona. Now David wondered how this beetle might behave in central New Mexico. Were the turrets much the same? Larger? Smaller? Closer together?

I didn't expect to find out in a single afternoon. In truth, I was taking a day of my life to drive to Laguna del Perro–Lake of the Dog–not for its tiger beetles so much as its name, just as other people are intrigued by the words Outlaw Spring or Hells Half Acre or Martha's Vineyard. I felt that linguistic pull. A mysterious resonance.

Laguna del Perro.

I also liked the pictures I had seen on the Internet: dry, flat, uninhabited land. In the West, this is the kind of marginal moonscape we associate with the Bureau of Land Management or someone's far-flung undevelopable ranch. Like the gal who finds the good in a bad man, I'm drawn to places with zero appeal.

In fact, the salt lake turns out to be extraordinarily beautiful. The effect is painterly: the lower portion of the scene a brushstroke of pale brown and dark brown clay-rich soil, with water spread thinly over that surface; a rising of tan bluffs in the foreground; low beige foothills in the distance; and an enormous cerulean sky. Against these minimal colors is a sparse scattering of grass, pale green and dark green. Blue and white, the lake reflects the cumulus clouds forming overhead like the chorus in a Greek play, setting the scene for the gods and heroes, the dark-grey, emotionally wrought cumulus-nimbus.

There's just no end to the view. Nothing to stop the eye. Nothing human. My husband and I walk to the water's edge, collector's nets in hand, collecting pounds of mud with every step. After an hour of slogging and no tiger beetles, we sludge and slurp toward the uplands where the ground is dry and rocky and where we stop to admire the occasional horned lizard, the way its riffled skin and pointy protrusions blend into the contours of sand and gravel. In my childhood we called these horny toads for their round bodies and blunt snouts, and I am still wary of handling these reptiles, knowing they can squirt a stream of blood from the corners of their eyes, a liquid that tastes foul to predators, stains clothing, and grosses out children. Horny toads also puff up when threatened and aim their cranial horns right at you, trying to make themselves look large and difficult to swallow. In short, these guys have a lot of personality.

Finally, atop a cream-colored bluff, among scattered cow pies, in a slough of salt grass, I see that movement in the air just above the ground–a glinting, dark-green tiger beetle. Getting out the guidebook, I tentatively identify it as the wetsalts tiger beetle or maybe the punctured tiger beetle or maybe the cow path tiger beetle. I'll never know since we can't catch the insect, not this one, and not any other. These beetles are supernaturally fast, darting into dips and vegetative tufts, slipping under the net–laughing at us. It becomes obvious that catching tiger beetles on the flat river sand of the Rio Grande has given us delusions. With the cow path tiger beetle, we are way out of our league.

Meanwhile, the view is singing hallelujah. The heart expands in the ribcage. The ribcage seems to soften. Thunderclouds billow and mass, catching light and throwing it back in shafts and halos, a mystical moment that keeps extending until the mind simply loses interest. The theatre

Laguna del Perro and Sharman Apt Russell (photo by Peter Russell)

of monsoon plays all summer long, and there's always another storm to applaud.

Peter finally mentions that word "storm." Although lightning kills relatively few people every year, being outside in a lightning storm and carrying a long aluminum pole would seem to increase the odds. By how much? I wonder. And how much did these aluminum-handled nets cost, anyway? Maybe we should dump them.

Rain is absurdly local in New Mexico. Thunderclouds billow and mass and burst, and it happens on your neighbor's house and not yours. We hurry back to the car and drive west toward home, and by the time we reach the Rio Grande again, where Highway 60 crosses the river, the afternoon sun is shining. We stop for one more look under the bridge. No tiger beetles. But this time, instead of staying near the water's edge, I pace the cliff bank, activating my search engine for those small holes that indicate a larval burrow.

The entrance to a tiger beetle's larval home ranges from less than a millimeter across to over ten times that, depending on the species and stage of the larva. The mouth of the tunnel is usually an almost perfect circle, flush to the ground, with a slight flattening on one side where the larva can rest its mandibles. Some species also create a shallow pit at the entrance into which prey more easily tumble, and the Williston's tiger beetle constructs those intriguing turrets. At the burrow entrance–its head and thorax dusted with the color and texture of the surrounding soil–the larva waits for any small thing that moves within striking distance. Enough meals and the hole is plugged while the grub sheds its skin, molting from first to second to third instar, and then enlarging its burrow to fit its new size.

Just down from the bridge, the bank of the Rio Grande is defined by a four-foot-high shelf of chalky soil. At the base, three circular holes stand out as if freshly drilled. Of course, a number of insects make holes similar to tiger beetles, particularly the burrowing wolf spider. Smooth round holes in sandstone bluffs are likely the burrows of mining bees. Some wasps and solitary bees also have burrows with round openings, although these tend to be irregular. The emergence of cicada nymphs can leave a circular hole, too, somewhat larger than most third-instar tiger beetles.

I look at these particular holes along the bank of the Rio Grande, so nearly round, so nearly perfect, and have a good feeling.

Peter and I still have a four-hour drive to the Gila Valley, and we consider the different methods of capturing tiger beetle larvae. Tunnel lengths range from six to seventy-eight inches. Most burrows go straight down, although a few species dig into the sides of cliff banks or the branches of rotted wood. "Fishing" for a larva involves inserting a blade of grass into the tunnel and hoping the grub will bite the stem and allow itself to be gently pulled up. In tunnel blocking, you wait for the larva to move to the top of its burrow and then insert a knife just below that point. The third and least delicate method is excavation, quick and dirty, and that's what we choose. Peter presses the tip of a shovel in front of the small hole, down a foot and then up with a mess of dirt, roots, and rock. The first two holes are empty, or perhaps we haven't reached the bottom of the tunnel. But in the third, crawling unhappily in this new, brightly-lit world, is a half-inch, instantly recognizable tiger beetle larva, undoubtedly a third-instar, monstrous with two back hooks and a heavy head dominated by slice and dice mouthparts. The body thickens at the section with the hooks and curls down, giving the larva a kind of rump, something that could easily sit on a stool in the bar scene from *Star Wars*. Grotesque! Beautiful! A horror! A prize! Contradictory emotions produce a kind of buzz in the head.

I put the grub in a jar of dirt from the same crumbly bank. Later I'll transfer it to a larval rearing tube of plastic piping. It's almost four o'clock, the sun shining, a phoebe singing from a willow tree. I check my ten adult beetles in their collecting boxes. Everyone's buckled in.

July 31

The next morning I seemed to have killed them all. Well, not the larva, and not four of the tiger beetles, which are scratching frantically at the glass of the terrarium like prisoners screaming let me out, let me out, let me out, let me out. But the remaining six adults are humped inertly in the sand, half-buried as if having been forced to dig their own graves. Moving the two collecting boxes away from the morning sun calms the four hyper adults; tiger beetles respond to light and will exhaust themselves trying to reach any source that comes from the side of the terrarium. And then, within an hour, all the tiger beetles in the terrarium rise up like the walking dead, their mandibles scything with anticipation. When I consult the guidebook again, I discover that at night, bronzed tiger beetles dig shallow burrows in the sand from which they emerge for breakfast.

Good morning, and time for mini-mealworms! The larval form of a darkling beetle, mealworms can be bought online in five sizes: mini, small, medium, large, and giant. Minus the handful eaten yesterday and the twenty or so spilled in the car, I still have a thousand mini-mealworms in a round, eight-inch-high cardboard container, enough to last the rest of my life. Although I serve them chilled from the refrigerator, the small brownish grubs start to move quickly, wakening to a new promise in the balmy summer air. They jerk, they squirm, and the adult tiger beetles pounce, grabbing their prey in warrior mandibles, the worms moving more seriously now, writhing and flailing. Two pairs of the bronzed tiger beetles are already mating, the smaller male riding the female's back, and only the female gets to eat, the male unwilling to give up his position. For the first time, I watch my captives closely and at length, fascinated and uneasy.

This is my job now, feeding this appetite, day after day. Like some ogre in a fairytale, from the cardboard can of a thousand mini-mealworms, I choose which mini-mealworm lives and which mini-mealworm dies. The birth of the universe began with a bang, and the banging has never stopped. The bloodbath of nature—who am I to object? But this deliberateness, selecting each grub with my tweezers and dropping the creature into the terrarium, feels decidedly awkward.

I think of Jain monks who cover their mouths to avoid ingesting any insect in the air, monks who walk barefoot and sweep the ground so as not to crush more small animals. Jains believe that every living being has a soul and every soul is the architect of its life and potentially divine, on its ascent to bliss through karma and rebirth. The first principle of Jainism is nonviolence to all these souls, an idea that seems quixotic, impossible to achieve, since even microscopic organisms must be protected. Somehow the Jains have decided to try in a complicated system that includes householders as well as monks, with some ten to twelve million Jains (all vegetarians) actively practicing their religion in India and elsewhere.

And here I stand, the anti-Jain. The new me.

I like being a citizen scientist who can pause the science to feel sorry for mealworms and for herself and even stop observing tiger beetles to google Jainism where I learn that Jains do not believe in a creator or deity responsible for the universe. Instead they share my belief, also shared by Albert Einstein, that we are all part of the dance of matter and energy. In Jain cosmology, this dance is eternal and cyclical and governed by natural rather than moral laws. Karma is less a force of good than a matter of physics. Attachment and aversion result in karmic bonding. Nonattachment and nonaversion release that bonding. For a moment, I believe I could be a Jain. I could devote myself to the physics of the soul, every moment of every day structured by ritual and restraint.

I imagine telling Peter. We'll have to change our lifestyle.

✳ ✳ ✳

I know that my tiger beetles are bronzed tiger beetles—each one, the Jains would say, with its own soul, on an ascent to bliss through karma and rebirth. This identification is based on their metallic blue-green underside and on last night's burrowing behavior. But if I want to identify other tiger beetle species, I need to learn to use the taxonomic key provided in *A Field Guide to Tiger Beetles of the United States and Canada*.

Taxonomic keys begin with a series of choices about the characteristics of a plant or animal, with the correct answer taking you to the next series of choices, and then the next, and then the next, and then the next, until you reach a final destination. It's as simple as playing Candyland.

The first set of criteria in the taxonomic key are:

1a: Front trochanters with one (rarely two) subapical setae, middle trochanters with or without such setae (fig 4.6A)–15
1b: Front trochanters without subapical setae, middle trochanters also without such setae (fig 4.6B)–2

This is a rockier start than I had anticipated. I look at user-friendly Figure 4.6 and see that trochanters are a small segment of leg just before the first long section of leg (the femur) and that setae is a technical name for hairs. Subapical means *below*. All the other words in the sentence are ones I recognize: *front, with, one, rarely, two, middle, or, without,* and *such.*

In this case, I already know what species I am identifying and can work backwards from that answer. To reach the bronzed tiger beetle, I need to choose 1a, which takes me to 15 and more choices:

15a: Very small, under 10 mm, red, red-brown to brown; trochanters of middle legs without long setae–16
15b: Small to medium, over 10 mm, trochanters of middle legs with one or rarely two long setae–18

The correct answer to that will send me to 18, and the next answers to 32, 65, 66, and 67, all of which involve closer examination of the beetle's labrum and thorax, antennae and frons and genae, determining whether that labrum is glabrous or the clypeus densely to sparsely covered with decumbent setae or the pronotum trapezoidal in shape. Lewis Carroll comes to mind. "Twas brillig," I murmur, "and the slythy toves did gire and gimble in the wabe. All mimsy were the borogroves …"

The entire process, which obviously could take many hours, ends at 69a or the bronzed tiger beetle, where the front maculation is usually complete and connected to or only slightly separated from the marginal line and "the pronotum is narrow with front about the same width as back (proportions four units long to five wide)."

Knowing all the answers in advance should make the work of examining and identifying this insect easier. The first step is to confirm 1a, which states that there is one (rarely two) hair on the front trochanters. Unfortunately, I can't find that hair on any of my tiger beetles. Not using my close-focusing binoculars to observe them through the glass, not using a hand-held magnifier, not using my reading glasses.

Peter thinks he can, peering into the terrarium, also trying to use the magnifier and then abandoning that for glaring through the plastic wall. "There, right there. Do you see it?"

"No. There?"

"No. *There.*"

"I don't see it."

"There! I think. I don't know. Do you see it?"

"I don't see it."

My immediate neighbor is the regional field director of The Nature Conservancy and has a good microscope that I am always free to use. Once I looked through this instrument at an oak leaf hopper—the psychedelic colors, the Art Deco design, the sheer weirdness—and literally gasped, a physical cliché I have performed only a few times in my life if you don't count giving birth. To observe insects, my neighbor freezes them first for ten minutes. Sometimes they survive the ordeal and sometimes not. Now I walk the little bit to her house with a sacrificial tiger beetle, willing to give it up for science or, at least, for my ability to identify tiger beetle species. This is pure attachment on my part, desperate to find that one (or rarely two) setae.

I've always had bad eyesight. In third grade, I couldn't see the big E when the nurse unveiled it on the other side of the room, and so I was diagnosed legally blind and given a pair of glasses, the sporty kind with decorative rhinestones. I wore eyeglasses and then contacts until someone invented laser vision surgery. In 1996, when I was forty-two, I got my eyes smoked. Smoke, my husband said—because he watched a video of the surgery—literally billowed up like a small brush fire from both corneas. I had ten good years until a few years ago when I needed glasses again, for reading this time.

I considered giving up reading.

Instead, I buy reading glasses and abuse them. I buy reading glasses and sit on them or put books on them or stuff them in my bag without their cloth eyeglass case. I buy reading glasses and refuse to wear them in a chain around my neck or pinned to my shirt. Yesterday, on the Rio Grande, I smashed my latest pair against my close-focusing binoculars and left them behind on some rock I will never see again. I have spares, of course, all equally scratched. I was born too flawed to survive in the natural world. As a hunter and gatherer, I would have died or been shamed as habitually incompetent. I sometimes think about this.

I just want to see those single hairs on the front trochanter. To get past step one in the taxonomic key. Under the microscope, the frozen bronzed tiger beetle is as spectacular as I had imagined it would be, an iridescent god–surreal, virile, powerful. The microsculpture on the wing covers is dramatically revealed, a new landscape, edges and dips, a geometry reflecting and refracting light so that round every turn is an explosion of color, glistening and gleaming. The mandibles are swords. The eyes are caverns. The legs are forested with Martian trees. You could worship this insect, just as the dung-rolling scarab beetle was worshipped in the Egyptian city of On for its symbolism of the sun being rolled across the sky. Just as the scarab's pupa became an inspiration for an entire culture of mummification in which humans, like beetles, could rise up metamorphosed, defeating death.

But in the confusion of shivery-god body parts, folded legs, and wayward antennae, I still don't see the defining seta. *Not even under the microscope.* I trudge back home with bad karma and an unconscious tiger beetle.

Along the length of my porch, where the ceiling connects to the house and wooden rafters, colonizing cliff swallows have built their protuberant nests: round mud baskets with a mud tube extending like an elephant seal's nose. Because of the dry spring and summer, a smaller insect population, and unknown ripple effects, this year we only have thirty swallow nests, or sixty swallow parents, and forty more swallow children, some still in their nests, the rest swooping and darting and cutting through the air. The birds surround me like a school of minnows, flashing above my head and to the side and behind me, and that is how I feel now in my pursuit of tiger beetles–like a woman swimming in deep water. A fish out of water. Out of my depth.

Storm over mountains (photo by Elroy Limmer)

August 2011

August 4

The rains have come, not every day, and not that much. Still, we stand marveling on the porch, inches from the satisfying curtain of water that will last twenty minutes. Or we watch the clouds gather and darken, the grey slant that means somewhere over there, over there someone is getting wet. Not us. But that's okay. Somewhere it is raining, water seeping into soil, following gravity and slope, finding its way to the Gila River. The river, by now, has long since flooded the banks where I watched the Western red-bellied tiger beetles eat the dead frog. They are gone, just as the guidebook predicted: "After the onset of the first significant rains there apparently is a dispersal event as this species' population numbers fall drastically in lowland areas just as individuals appear in adjacent higher altitude areas."

The authors of that guide were looking at the Western red-bellied tiger beetle in southern Arizona. But I am at a higher altitude, over forty-five hundred feet, and although the tiger beetles along the Gila River have dispersed completely because of flooding, those at nearby Bill Evans Lake have not. I stop at the lake, which is really a large pond, and watch the beetles skirt the edge of water, less a group now than individuals spaced carefully apart like humans on a crowded beach. Today the insects seem more interested in hunting than mating, skittering across the sand, stopping for long moments, seeming to think–what would I give now for a dead frog.

Of course, this is shameless anthropomorphization, ascribing human qualities to nonhuman things. And I do feel a little ashamed, out of habit and cultural training. In truth, though, I am no longer convinced that ascribing human qualities to nonhuman things is at all bad. Maybe we don't ascribe them often enough.

In his book *The Forest Unseen*, biologist David Haskell observes a snail and wonders what it is seeing. Experiments show that snails respond to black dots on a white test card and know the difference between gray and checkered cards. David wonders if snails see as we do, "with images of checkered cards appearing in their gastropod minds? Do they experience private displays of light and dark, processed by tangles of nerves into decisions, preferences, and meaning?" The cinematography would naturally be different, more film noir than not, more avant-garde than human realism. But a similar experience, even so, based on nerve structure and chemistry.

Our cultural story is that this snail movie "plays to an empty house." There's no observer in the theater watching the screen; there's not even a screen, but only light from the eye's projector stimulating certain wiring to move and eat and mate. In this story, the world is populated by billions of such "hollow theatres" lumbering about a bit monstrously–except for us, of course, *Homo sapiens*, whose consciousness has reached some kind of gestalt or mega-order of sentience incomparable to other animals.

Eco-philosopher David Abram argues against that elitism. His animate world is made up of sentient creatures rather more like us, a world of fellow beings and constant greeting. Hi. Hello. How are you? Hi. Grunt. Nod. Good morning. Hello. Even the inanimate Earth has some form of consciousness, part of the larger commonwealth. In *Becoming Animal*, Abram quotes a Mattole Indian, native to California, "The water watches you and has a definite attitude, favorable or otherwise." In Mattole culture, humans are advised not to speak before a wave breaks, not to speak to rough water in a stream, not to look at water for very long unless you have been to that spot ten times. From my modern perspective, the etiquette seems obviously sound–an unwillingness to treat the watershed carelessly.

Here in New Mexico, administrators at the state level want to divert the Gila River into a concrete pipeline that would carry the water to a city like Deming and then perhaps Las Cruces or the Rio Grande to be emptied into that over-adjudicated canal as part of New Mexico's water settlement agreement with Texas. They want to cut up and dismember the last free-flowing river in New Mexico, moving its dead body elsewhere. This idea is being considered quite seriously and has been done, in fact, to other rivers around the world.

I have been told all my life to think of the natural world as mechanical, not at all like me. To think otherwise is to think like a Mattole Indian, sweet and naïve. Of course, now many of us are wondering: who, exactly, is being naive?

* * *

Today I am driving on to the Gila River Bird Refuge, the dead-end of a dirt road past hackberry thickets and cottonwoods and oaks. In my experience, this is the best place along the river to see coatis. Relatives of the raccoon, about the size of a big house-cat, coatis have long noses; long, banded tails; a complex social and matriarchal structure; and conversational sounds– grunts, chitters, churrs. They project tough love (the bachelor males having to leave the group when they reach sexual maturity) and tough-nuts-to-you (staring balefully down from the treetops) with muscular shoulders and thick pelts in a range of brown, from black to golden to red. Their dark faces have patches of white that make them look masked; the versatile, useful nose is also white. Like other species affected by climate change, the white-nosed coati is steadily shifting to new habitat, in this case moving northward.

Driving slowly, I hope to see the band of mothers, aunties, and juveniles that now live here, where I live. Peering into the hackberries, I dream about studying coatis like Jane Goodall studied chimpanzees. (You could describe Jane Goodall as an early citizen scientist, a young woman who followed her passion and who was first scorned for that by the scientific

Gila River (photo by Dennis Weller)

community. On impulse, I decide to e-mail her, asking what she thinks of citizen science in the twenty-first century.) Watching coatis would be a different kind of project, requiring me to quit my teaching job and leave my husband–or perhaps he would quit his job and come with me?–to join this particular matriarchal band, living with them day after day, camping out on the Gila River, traveling with coatis from forest to forest, burrowing vertically into knowledge.

Vertical would be a new direction for me since, like many people, my understanding of the world is almost completely horizontal. I know a little bit about a lot. I stretch around the world knowing a little bit about local affairs, state politics, the scandals in the White House, ocean chemistry, scarab beetles. I don't go deep. I don't tunnel down into soil and root, frost and mineral, seasonal patterns, parasite loads, kinship bonds.

If I lived with the coatis, I'd build a tree house and sleep under the stars every night.

The path from the dirt road to the river dodges at last through shaded willow, that birth canal entrance into water and sun, the explosion of light. I'm humming along the bank with my long net and collecting boxes, looking for tiger beetles and thinking about my life. About my failures. How I didn't do this, didn't do that. I surprise three ducks, green-headed mallards, and they fly away in a triangle, quacking, and I feel that giggle like the child always amused by peek-a-boo, never getting tired of the joke: ducks actually quack, *quaack, quaack*, complaining and petulant. The delight of onomatopoeia. The delight of remembering that word. A toad hops on my foot, a little bit of moving mud. Water rushes over rock. The fruity smell of decay. A butterfly sailing past like a hot air balloon. An American painted lady.

Every few feet, I spot another Western red-bellied tiger beetle, that pattern of seven blotches with a musical note. Then one of the beetles has a different pattern, four creamy dots, and I'm not only looking at an ocellated tiger beetle, I *know* I'm looking at an ocellated tiger beetle. David Pearson calls this species the house finch of tiger beetles because it is so common in the Southwest, abundant around water edges, more solitary in the uplands. Sometimes these beetles climb shrubs and plants to roost at night or to escape the hot surface soil midday. I watch this particular beetle until it is approached by another ocellated, and when they both fly up, I can see the reddish ends of their abdomens signaling (like the Western red-bellied tiger beetle) that they might emit poisonous chemicals and be bad to eat. The rest of the abdomen is metallic dark-green.

Pattern recognition. Four creamy dots. Something in the world and something in my brain snap into place like the two ends of a Tinker Toy. Tiger beetle and butterfly enthusiasts share this satisfaction—matching up beauty with order. Chevrons, bands, circles, dots. We can identify some species fairly quickly. The mind has a picture. The picture matches. The square peg goes into the square hole. Birders also know that pleasure, a flash of color and form, and they have an answer: scrub jay or vermilion flycatcher. This kind of competence feels completely right. It's as if we belonged here, as if nature were our real home.

I confess to that Paleolithic nostalgia. We are hardwired for walking through the woods, along the river, feeling at home, matching patterns, knowing what we see and what to do next. Willow and bear. Mountain lion and squirrel. They make sense. They may even feel like family. Hello, good afternoon! Relatives that are friendly, and relatives that are dangerous, people we have known all our lives. We've replaced these competencies with new ones. Books and computers. How to use a toaster oven. Perfectly reasonable, I tell myself, as I look for tiger beetles along the riverbank, as I think of my achievements—the machines I know how to use, the machines I don't know how to use, the machines I know how to use but don't know how to fix—as it seems to me suddenly that we've replaced these competencies with a thousand incompetencies, that I live in a world I understand less every day.

Something darts by my feet. Another tiger beetle. And I'm focused again. Another ocellated.

I eat my peanut-butter sandwich on a slab of rock beside the Gila River, birds singing madly in the cottonwood trees, water rippling, insects humming. Twenty years ago, Peter and I brought our children to this spot to watch them play. Of course, I knew how to love my children. I understood that fierce all-embracing love, and even today I know this love means your children leave you. Toss you aside. Perfectly natural, I tell myself. I remember David and Maria with a longing that can come on me at any time, an emotion I've learned to simply watch, those beautiful children, so happy to be with us, playing in the river. How short a time, how halcyon, that part of my life.

Throughout the day, a long, complex afternoon, I can think about my life because I do not have to think about my dinner. I won't go hungry tonight if I don't find a deer or catch a fish or dig up some roots. Hunting and gathering, the Paleoterrific, has its pros and cons. I'll never know what they are since imagining that life is only a dream of received ideas, stories we make up from scanty evidence, dreams built on dreams—a sense of loss, a vague excitement.

The best part of this day is when I focus on tiger beetles and the insect drama at my feet. But a good part of the day is when I look at my life from a certain distance and feel pleasantly relaxed about all that. A good part of the day is when I think about what it means to be human, forty thousand years ago and right now in the twenty-first century, when I let myself range across time and space, one of the stranger competencies of the human mind.

August 5

In the middle of the night I wake up worried about the Rio Grande tiger beetle. I'm on a quest to know more than anyone else on the planet about this species, but first I have to find the insect. In the middle of the night, the psyche has the equivalent of a sugar low. There's no resistance to doubt. In the sunshine, I'm a problem-solver. In the middle of the night, I'm the little engine that can't. The shadows are bigger and the mind speaks some archaic language of doom.

Morning, and I go back to the guidebook to discover that the larval biology of the Western red-belled tiger beetle is also "unknown in the wild." What does that mean? I e-mail David Pearson and an hour later he has e-mailed back: "Yes, and this is a paradox with such a common species as *C. sedecimpunctata*. We tried to raise larvae with mating pairs in a terrarium, but the females wouldn't oviposit. This is a method we used successfully with several other species. We think the adults may migrate to higher altitudes after the rains to oviposit somewhere there, but who knows?"

David sends this out like a message in a bottle. This may explain why amateurs, people like me who pick up bottles, ever hopeful, write 80 percent of papers on tiger beetles.

In the afternoon, I drive back to Bill Evans Lake and collect ten Western red-bellied tiger beetles, putting five of them in a large terrarium with sand from the water's edge and five in a terrarium with soil from the raised cliff bank rimming the lake but away from the water. The Western red-bellied tiger beetle has become my back-up plan.

* * *

At dusk, around seven in the evening, Peter sees a mountain lion crossing the bridge over the irrigation ditch, about a hundred feet from our house. At first he thinks the animal might be a fox or coyote, but the body is so much larger and the turning head is distinctly leonine. There's no mistaking that round shape and flat face and long tail flowing past. Peter says he has never seen anything so alien, the expression without noticeable curiosity or fear. (Actually, Peter says he was reminded of the *Terminator* movies when the Arnold robot assesses the world, the little screen in that head ticking out facts: height, weight, distance, prey?)

Over the years mountain lions have left their prints up and down the irrigation ditch, on the dirt road to our house, and on nearby paths by the river. Studies show that mountain lions adapt well to moving through human environments, along ditches, into culverts, sidling through rose bushes. More and more people in the suburbs of Silver City, for example, have seen the scrapes and scat of lions as these neighborhoods attract more and more deer. One friend heard the death cries of a doe in the night and found its eviscerated corpse in the morning, close to her bird feeder.

I've lived almost fifty years in the Southwest and have never seen a mountain lion in the wild. I've also lived with Peter for thirty-three years, so it really counts that if he has seen one, I have, too.

August 11

I collect ten more tiger beetles at Bill Evans Lake, as well as buckets of dirt from the steep hill just above the lake, digging for soil at the base of clumps of grass. Before laying eggs, females use their mandibles and antennae to test for texture, salinity, moisture, and temperature. Different species also choose sites based on the size and shape of their ovipositor or egg-laying organ. Research suggests that in desert grasslands, females lay more eggs on the eastern side of plants, where the soil is warm in the morning and shaded in the afternoon. Because this beetles' habitat is likely to flood after the rains and because they are known to disperse, my prediction is they will prefer this upland soil to the sand by the water's edge or the rockier cliff banks nearby. From a mother's perspective, I like the upland soil too, which is crumbly and soft and rich with organic debris.

In the first three days of captivity, from August 5 to 7, my Western red-bellied tiger beetles were not mating, quite unlike the bronzed tiger beetles who mated almost immediately and have continued ever since. A male tiger beetle courts a female tiger beetle as if she were something to eat. After running down his prey, however, instead of stabbing and jabbing, he leaps on her back, grips the sides of her neck with his mandibles, and grabs her wing covers with his middle and front legs. The female tries to throw him off. The two struggle and stagger as the male inserts his extended genitalia into her genitalia, introducing his sperm into her sperm storage compartment. Once successful, he may stay attached for hours, like some nightmarish backpack.

A female expends a lot of energy producing and laying eggs. In bio-economic terms, she wants to get the most from her investment. Her violent response is likely a way to test the male's strength and endurance, and to ensure that the fitting parts are the right ones, preventing other species from mating with her. (Evolution doesn't personalize like this, of course. Rather, females who don't struggle or test their mates have less luck over time passing on their genes, and so the trait of the passive female dies out.)

Males probably continue to guard the female so that their sperm will be the last sperm released to fertilize her eggs. In terrariums, at least, males ride females longer in the presence of a second lone male. David Pearson and Barry Knisley also tested the hypothesis that piggyback males were protecting the carrier of their genes from predators, but the two men found that they could actually catch a guarded female more easily than a solitary one—as, presumably, could lizards and birds.

In an effort to get my Western red-bellied tiger beetles to mate, I tried a darker environment, which the bronzed tiger beetles had seemed to like or at least tolerate. Then I tried a hotter environment, bringing my desk lamps closer to the terrarium and using higher-watt bulbs. I googled papers and blogs on the Internet. In 1984, Barry Knisley reported that the Eastern beach tiger beetle didn't lay eggs until the temperature reached above 82 degrees Fahrenheit–and stayed that way at night when the females did most of the laying. A site run by the University of Massachusetts recommended an "environmental control chamber with a 12/12 or 14/10 hour photoperiod with a cycle temperature ... warming concurrently with onset of light, gradually climbing to 28–29 Celsius in the middle of the photoperiod and descending to maintain a low of 18–17 Celsius for the duration of the dark period." I went to the website BugGuide.net and contacted the well-known beetle blogger Ted MacRae at Beetles in the Bush. Overall, the advice was patchy, one thing for one species and one for another. But within the confusion, I also found a surprising warmth, even friendship. Here we are together doing this unworldly thing no one else cares to do. How can we help each other?

After two more days, in temperatures kept at 92 degrees Fahrenheit for five hours a day, the Western red-bellied tiger beetles began to pair off until each terrarium had at least one female ridden by a male. Maybe it was the heat or maybe something else. Unfortunately, I didn't use a control group, one without extra heat. And my desk-lamp protocol lacked consistency since the electric timers I bought almost immediately stopped working, so that instead I turned the lamps on and off manually, whenever I recalled I was doing this interesting experiment that involved me paying attention.

Today, after lunch, I put the ten beetles just collected at Bill Evans Lake under the same regime, two groups of five in two more terrariums with grassland soil, a cheap desk lamp set close overhead. Puttering among the equipment–my thermostats, my GPS, my recently purchased digital calibrator–I remember to look at the insects as well, and I see something new among the other Western red-bellied tiger beetles who have been mating for a few days now.

When females lay eggs or oviposit, they often rear back almost vertically and make thrusting movements with their abdomen, digging or cutting with their ovipositor into the soil. At the right depth for their species, they lay a single oval-shaped egg covered with a sticky substance to keep it in place. (Species differ, of course, and a few tiger beetles lay eggs much deeper in underground burrows.)

Right now, in my large terrarium with soil from the cliff bank near the lake, a female still ridden by a smaller male is standing in the corner, straight up as though delivering a speech before an invisible lectern. As though crossing the Delaware. As though looking out like Scarlett in *Gone With the Wind,* thinking, "What is there that matters? Tara! Home. I'll go home!"

I look for thrusting movements of the abdomen and pretend to see one. Soon I am e-mailing David with the news.

August 12

> From: David Pearson
> Sent: Fri, 12 Aug 2011 1:39:13 PM
> Subject: RE: laying eggs
> Great! The trick is to keep the soil moist but not too moist. After they hatch out, larvae will come to the top of their burrows and actually "learn" to take small insects from forceps. The male can be removed soon, and then let the female oviposit her eggs. In a week or so, you can remove her, too. Wow. That's great. You may be the first to see and describe the larval characteristics of this species. Dave

Great, yes. But also a great burden. Now I worry about the eggs, afraid they will get trampled with all the activity in the terrarium, two mated pairs and one lone male. I wonder if I should take the other tiger beetles out, but I'm afraid they might escape in the transfer and I don't have any more terrariums prepared.

Instead I pack up the car because tomorrow I am scheduled to go back to the Rio Grande to look again for the Rio Grande tiger beetle. I finish commenting on student papers in my online summer school graduate classes, Advanced Creative Writing and Writing for Social Change. The students post in discussions on alliteration, plot twists, the omniscient point of view, sex slaves, minimum wage, and cruelty to animals. I e-mail the Jane Goodall Institute again: did she get my last message? Is she excited by the new world of citizen science? Does she wake up every morning filled with hope, if only tenuous and dawning, for the potential of our species? My virtual life is rich and complex. My tiger beetles are laying eggs.

August 13 and 14

Caballo Lake State Park is a recreation area just south of the Bosque del Apache National Wildlife Refuge—one of seven dams and reservoirs on the Rio Grande. The 11,500-acre lake is gray and calm under cloudy skies, the 135 camping sites mostly empty, and only a few boats chug the unappealing water. The shoreline is rocky with small dead things lodged in crevices—washed-up fish parts and candy wrappers. I walk the lake edge, checking spots of mud and sand for tiger beetles, then turn a corner that leads back to a lagoon, its flat surface dried, patchy with scrub grass, littered with red bobbers and metal sinkers. This is promising tiger beetle area, a kind of NASCAR racetrack for the predator to build up speed and outrun its victim. I look for the telltale flash and dart but am disappointed. Along the gravelly bank, walking faster, I see an earring that someone has lost, glittering between rocks. The jewelry moves deeper into earth. I pause to push the stones aside and the earring skitters up and away, squirming into another pile of gravel. Not a tiger beetle, I think, with that horror of sunlight. And probably not an animated earring. This is one of those experiences when you simultaneously doubt what you see and hope that you have entered into some private Narnia or possibly *The Matrix*.

So I move more rocks and expose the reluctant, fantastical creature, an inch long, its body blue, green, and red, the colors swirling like an oil spill. With heavy mandibles, big eyes, six long, pale legs, and two pale antennae, this is, in fact, a tiger beetle, the bottom of its iridescent wing covers decorated with two crescents facing each other, the rest of the wings pitted, ridged, and undulating in a microsculpture covered with pigments and waxes that reflect light.

I had forgotten about the three genus of nocturnal tiger beetles: the giant tiger beetles, the night-stalking tiger beetles, and the big-headed tiger beetles, their greatest diversity in Bolivia and Brazil but also appearing here in the southern United States. I wonder briefly if this could be a rare species, something to make David Pearson gasp. I am too surprised to actually collect the beetle.

Which disappears. But now my collecting box is out and I am walking the gravelly bank where I see another gleam, which also squirms into rock. Squatting, I look at the mud flat, narrow-eyed like an art dealer examining a painting, and there is yet another glitter, at some distance, and I creep toward it with my long-handled net. Unexpectedly I'm in the flow and slam, hop, I catch the insect and hurry back to the empty parking lot and my car and guidebook. Although the color plate shows this beetle as dull black, the text says differently–"metallic maroon, green, and purple upper surfaces." I'm not sorry to learn that this phantasmagoria, the Pan American big-headed tiger beetle, is fairly common, ranging from California to Florida. Mostly nocturnal, and found near water, the beetles occasionally come out on warm cloudy days. Sometimes people see them under artificial lights, at a campground restroom or gas station or blinking store sign. I never have, of course. I didn't know to look.

The Rio Grande is another surprise. The river is almost completely dry, without a single braid or channel of water, only disconnected shallow pools. I see a scattering of bronzed tiger beetles at one of these pools and release the captives I have brought from home. The riverbed has become a canvas, pleasingly repeated ripples and waves left by departing water, with animal tracks in patterns of hunting and eating, drinking and fleeing. The mud has cracked into large irregular tiles, their dried edges curling up, which make a satisfying crack when stepped on.

Deep in the crevice of one mud tile, I see another iridescent gleam— another Pan American big-headed tiger beetle. It's a lesson in spatial perspective. Fall in love with mountain lions and you will be lucky if your spouse sees one after fifty years of living among them. Fall in love with the Pan American big-headed tiger beetle and you will find yourself dazzled, requited love, at every turn.

But no Rio Grande tiger beetles. I drive to Socorro, a small town on the freeway twenty minutes north, get a hotel room, and then continue to the bridge at Highway 60 and the Rio Grande River. Since this tiger beetle is attracted to lights at night, I'm going to use my fluorescent lantern and a white sheet placed on the ground. In such circumstances, catching beetles by hand is often more practical than using a net. In his book on the ecology of tiger beetles, David Pearson explains:

> If you cannot use the tips of your fingers, try placing your
> shallowly cupped hand over them. If you do not cup your hand,
> you will likely flatten the specimen. Feel for the movement of
> the struggling beetle and then move your hand back so that your

Tracks in the Rio Grande mud (photo by Sharman Apt Russell)

closed fingers are over the beetle. Carefully spread your fingers
just enough so that the beetle runs up between two fingers to
escape. At that point, close your fingers to capture it.

I've read this passage more than once, and I still miss the trick. Also, I
don't want to be bitten by a tiger beetle; I handle them exclusively through
the cloth of the net. Reality check: the mandibles of the quarter-inch-long
Rio Grande tiger beetle are small. It's not the puncture or pain I fear but
the startle and intimacy of the bite.

And now I'm on the river, a stretch of water about eight feet across,
flowing modestly under the bridge of Highway 60. The sky is no longer
cloudy but lit by a full moon, that almost perfect circle, a floating face
in the sky as strange tonight as any night since moon first orbited Earth.
What holds up the moon? What thread links me to this circle of light? Is
she my real mother? The air smells of willow and mud. Occasionally on the
highway bridge above me and to my left, a car rushes past.

When I got here at dusk, bronzed tiger beetles were still idling at the
water's edge. Now they are burrowed into sand. My fluorescent lantern

isn't attracting many insects–a few moths, a few mayflies, a few non-tiger-beetle beetles–and the scene feels desultory. The moon turns us inside out, pins us to the ground, and then floats there unconcerned.

An hour goes by.

Ten minutes.

Another ten.

My Honda Civic is parked on the highway shoulder, a hundred feet away, and suddenly I see the headlights of someone stopping near the bridge. A door slams before the car drives on. I'm thirty-five miles south of Belen, another small freeway town fifteen miles south of Albuquerque. I don't know why Belen has such a high rate of violent crime, higher in 2009 by over 40 percent than the state of New Mexico and over 100 percent than the nation, with a property crime rate of more than 150 percent the national average. I don't know if someone just stopped at my car and then left or if someone is up there now, on the bridge, on the highway.

I think it's prudent to wait. To see if anyone I don't want to know drives off in my car. I listen as well for the sound of someone making his way toward me down the path from the highway. That path is full of rustling plants and slippery spots, particularly under the bridge, which is low enough in places that you have to crouch. Earlier in the evening I had hit my head hard enough to raise a bruise. It's not an easy scramble to do in the dark, even with a flashlight.

But there is no light, no sound, no one here but us moths and mayflies. I look at the white sheet on the ground dotted with a half-dozen insects. Science is full of such moments. Tedium. False alarm. False hope. I feel a prophecy coming on, rising up in my throat, breaking forth from my lips: the Rio Grande tiger beetle is a no-show. It's just another spectacular night of moonshine, the call and response of Earth and sky.

"This isn't working," I say out loud. Now I'm talking to myself.

Gathering up the fluorescent lantern, trundling up the sheet, back at the hotel, I'm conflicted. I like hotels. I like mid-range hotel rooms that have a coffee machine and microwave and refrigerator and TV and everything you need to stay here forever, a ship at sea, a little box that feels safe and familiar and reminds me of my childhood growing up in apartment buildings in the suburbs of Phoenix. True: the potential of the human spirit immediately dies in such a room. I could be outside now, lost in grandeur.

The next morning I make up for being a bad entomologist/person by taking the back road home to look for tiger beetles along a drying marsh, where I stay longer than I want to despite the mosquitoes. I sing along with Johnny Cash to "A Boy Named Sue." I park at a trailhead leading steeply to the kind of creek that in the Southwest we call the San Francisco River. This canyon area is lush, remote, humming along without human inter-ference–no dams, no reservoirs, no diversion, no agriculture. Spangly cottonwoods shade the bank, the water cold and clear and fast enough for native Gila trout. A dart and blur. The reunion is sweet. A Western red-bellied tiger beetle.

August 15

The tiger beetles at Bill Evans Lake are almost gone. After searching and walking the lake twice, I collect ten more–this time, two pair mate imme-diately, still in the collecting box–and now I have over thirty beetles in an office packed with plastic terrariums in different sizes, on stacks of books to get them closer to the desk lamps clamped to the back of metal folding chairs. More terrariums come in the mail today and I build a tower of the empty plastic boxes with their bright blue and green lids. It's a child's city. I move between the narrow aisles, parceling out mini-mealworms.

August 25

Today is the first day of my class called How to Become a Leading World Authority. The title is meant to be wry, a reference to Dick Vane-Wright's comment "You could spend a week studying some obscure insect and you would then know more than anyone else on the planet." Most (but not all) people get the joke, which I explain in my course description for the Western Institute of Lifelong Learning, a community group offering classes to students mostly over sixty and retired. I'm hoping for a room of wanna-be citizen scientists, some with projects to go, and others eager to pick projects from the list I'll provide. I've lined up local citizen scientists as speakers and am ready to facilitate an environment of support in the crucible of baby-boomer energy.

But only three people show up, and we immediately lack a certain mass as well as the shared belief in our popularity. Doubt grows from a seed in my own heart. If no one else came, why am I here?

Still, I'm the teacher. I have to be game. After introductions, I set a brisk pace, explaining my work with tiger beetles and the new options offered to citizen scientists. First–I mime excitement–go to the website SciStarter, a user-friendly and wonderful clearinghouse for citizen science projects large and small. Choose what discipline you want to work in, from archaeology to physics, and where you want to work, from "at the beach" to "in the snow." Click *search* and see what happens.

At the Mastodon Matrix Project, sift through the material surrounding an excavated mastodon looking for bits of bone, plant, shell, and rock that will help determine the environment where the animal lived and died. The packets of soil are sent to you by mail. You'll get instructions.

For the Great Sunflower project, plant sunflowers and watch what bees come to pollinate, helping create the first map of bee activity in North America, something particularly needed with the recent collapse of bee populations.

If you live part of the year in northern California, Oregon, Washington, or Alaska, walk one of three hundred beaches being monitored by the Coastal Observation and Seabird Survey Team (COASST). You'll be documenting the presence of marine animals.

A good online program is Stardust@Home where you can search through a million images scanned from a NASA spacecraft brought back to Earth in 2006. The cracked, uneven surface of the craft's collector contains an estimated forty-five interstellar dust particles. You'll be sent photos to study. If you spot one of the particles, you get to name it.

And if you are interested in something more grassroots, try the Albedo Project, which was started by a high school science teacher and her students. Its website explains: "Wherever you are–anywhere in the world–contribute to science by taking a photo of a blank white piece of paper!" Photos are needed on specific dates and you are assured that "your photo will be used to measure how much of the sun's energy is reflected back from Earth–our planet's albedo." You'll help discover how much heat different parts of Earth absorb. Is urban LA hotter than a desert in Africa? Can you guess what is the *hottest* spot on the planet? You immediately recognize the tone. Many citizen scientist projects are geared to educate the general public, beginning with schoolchildren. This sounds like something an adult would think fun for a sixth-grade class, who would not think it fun–but who might change their minds.

As well as joining organized projects, you can organize your own.

DIYBIO (do-it-yourself biology) is a network of amateur biologists, supported by biological engineers who design cheap lab equipment. You can use their centrifuges fabricated with a 3-D printer and learn how to extract DNA with kitchen appliances and a whiskey shot glass. The organization also sponsors public labs in Brooklyn, Boston, and San Francisco, where citizen scientists can practice their skills in synthetic biology.

Other self-organized projects would include studying the Western red-bellied tiger beetle in southern New Mexico, filling in that blank spot on a map.

Grab hold of a question and follow its thread.

I go through my speech and look hopefully at the crowd of three. Did anyone bring an idea to discuss?

One man wants to look at the history of religion. A broad topic and not science. Ever professorial, I say, "Okay?" and point him to some research tools.

But Ed Greenberg, who has recently moved to the small town of Silver City, New Mexico, from Washington, DC, leans back in his chair and lays out a rather nice plan. He's read an article on how to dehydrate green leafy plants into a concentrate high in protein, calcium, vitamin A, and iron. The plants range from cassava to kale, the processing is low-tech, and growing leafy plants for concentrate uses less energy for more protein and nutrition than any other agricultural crop. A main disadvantage is that these plants require a steady supply of water, often not available to poor farmers in dry areas. Ed has some experience with hydroponics, and he wonders if its water-saving methods could be combined with the benefits of leaf concentrate to help feed the world's one-in-four malnourished children. Ed likes tools and machinery and how things get done; he's directing that energy now to applied science.

And Allison Boyd, who used to be a paralegal working in real estate in Phoenix, has just become acquainted with Nature's Notebook on the website for the USA National Phenology Network. Phenology is the study of a species' life cycle or changes through time, and phenology is hot in these days of global warming as we contemplate a new timetable of plant and animal behavior. Allison is now keeping a "notebook" on what's happening in her backyard. She has a list of grasses and trees. She's enamored with the bumpy bark of alligator juniper. She tries, grinning, to remember the differences between blue grama and black grama and sideoats grama grass.

Feeding the hungry. Documenting climate change. We lean back in our chairs. What we lack in mass we make up for in gravitas.

August 30

I've failed for lack of a mister. I asked the guy at the hardware store, "Do you have any misters?" and he pointed me to the brooms. Wal-Mart is too scary. So I stopped trying, even though I really needed something to delicately spray the terrariums rather than pouring the water from a cup slowly, slowly, slowly before growing impatient and spilling out a great lake, drowning all my eggs and making the adults skitter and scream soundlessly.

Moist but not too moist. What does that mean?

Also, it's hard to clean terrariums with active tiger beetles inside, and some of the uneaten mealworms are starting to decay. I worry about

Terrariums in office (photo by Sharman Apt Russell)

fungus. Is that fungus? What is fungus? Not a plant, not an animal, not a bacteria. But its own kingdom.

Nineteen days have passed since I saw the female lay her eggs, and nothing has hatched. My terrariums look like hotel rooms no one has vacuumed in a long time. I've failed, but I am not surprised. David Pearson couldn't rear this species, and he knew what he was doing. Science is about expertise, and that's a comfort. Although not true. Alexander Fleming accidentally contaminated cultures of staphylococci in his untidy lab and so discovered penicillin. Watching the trial explosion of the first atomic bomb, some of the physicists wondered if it was going to ignite Earth's atmosphere. Science is about ignorance and failure and the possibility of failure, and I'm climbing this learning curve in baby steps, stopping often to admire the view.

And look at that. In just a few minutes now, I've negotiated the shift from failure to optimism. Isn't this, really, the story of my life? By God, I promise myself, I'll go back to the hardware store. I'll learn more about fungus. I'll get out that GPS and learn how to use it.

Western red-bellied tiger beetles mating (photo by Mike Lewinski)

September 2011

September 2

My friend Shirley comes with me on my last trip this season to the Bosque del Apache Wildlife Refuge. We stop at the visitor's center to talk with the biologist, who is young and blond and makes me think, briefly, that I'm in a television show where professionals always look this beautiful. We theorize about why we are seeing fewer species of tiger beetles than ten years ago in Barry Knisley's study. There's the ten-year drought, of course, and last winter's unusual cold spell. The biologist says those harsh freezes also killed a number of Western diamond-backed rattlesnakes living in metal culverts and that many of the white-winged doves overwintering here had frozen toes and beaks, which led to gangrene. A few oryx are missing ears due to frostbite. (Five-hundred-pound gazelles from Africa, oryx were introduced into New Mexico in the 1960s and are now overrunning their range.) A lot of prickly pear cactus died, and the screwbean mesquites were a month behind schedule budding out. We contemplate those maimed doves and oryx. The biologist tells me she hasn't yet collared the female mountain lion who birthed two cubs last spring. I give her back the bear spray since the visitor's center will be closed by the time Shirley and I return from the river.

With the biologist, I am careful to use scientific names, *Cicindela sperata* for the Rio Grande tiger beetle and *Cicindela repanda* for the bronzed tiger beetle. The reasons for a standardized system are obvious. The Eastern red-bellied tiger beetle, or *Cicindela rufiventris,* is completely unrelated to the Western red-bellied tiger beetle, or *Cicindela sedecimpunctata.* Similarly, the common name *stink-bug* can refer to *Halyomorpha halys,* an agricultural pest from Asia that has recently invaded the United States, or to the darkling beetle *Eleodes obscurus*, a comical insect native to the Southwest who scares predators by standing on its head and emitting a foul odor.

Moreover, common names are often whimsical, even silly, and science prefers the more somber mantle of Carl Linnaeus, who invented modern taxonomy in the early eighteenth century and who referred to himself in the third person: "God creates. Linnaeus arranges." When Linnaeus came up with a decoder ring that used the Latin of the ancient Romans–founders of Western civilization–his system was embraced almost overnight. One eighteenth-century scientist confessed that when speaking to his father as a boy he was *only* allowed to use Latin. In this way he learned that tongue before his native Swedish.

"Help, Father! I'm drowning!"

"*Filius, filius, lingua latinadicte!*"

Cicindela sperata. Cicindela repanda. Cicindela sedecimpunctata. The first word is the genus; the second is often but not always descriptive. *Repanda* for bronzed. *Rufiventris* for the orange underside of the beetle. *Sperata* for hope. (I'm not making that up.) The Western red-bellied tiger beetle, or *Cicindela sedicempunctata,* in the genus *Cicindela,* in the sub-genus *Cicindelidia,* in the family *Cicindelidae,* in the suborder *Adephaga,* in the order *Coleoptera.* The words clattering in your mouth soon become smooth as rocks tumbled in a stream. So quickly, you become part of the elite. *Tetracha carolina* in the genus *Tetracha* (which some experts dispute, putting the Pan-American big-headed tiger beetle in the pan-tropical genus *Megacephala*), in the family *Cicindelidae,* in the suborder *Adephaga,* in the order *Coleoptera.*

A *Rosa kordesii* by this name smells sweeter.

Back on the Rio Grande, Shirley looks at me sideways, and I stop. We are alert for the Rio Grande tiger beetle *(Cicindela sperata),* maybe the metallic green of an aridland tiger beetle (*Cicindela marutha*) or the fine patterning of a thin-lined tiger beetle (*Cicindela tenuisignata*). The river flows modestly with braids of thigh-deep water separated by strips of white sand and mud. For those of you thinking about traveling, September is the most beautiful month in New Mexico, the summer wildflowers still blooming, the air moving toward crisp, the skies brilliant blue with the occasional white cloud sharply edged and spiritually white. The bony brown mountains pile up on the horizon. Friendly Watchers. Hello. Hi. Good morning. They *do* feel like family. This *could* be the Elysian Fields. Water and sun, warmth and light: do I really need anything more? Do I need anything more than a raven caw-cawing and swooping down to a cottonwood by the river's edge? The big black bird produces a liquid, throaty gurgle. He seems to know us. Our lives seem special.

After a few hours, we still haven't seen any Rio Grande tiger beetles, just a lot more bronzed and one lonely punctured tiger beetle on the grassland above the bank, its wing covers dark and unspotted.

Devastated, Shirley and I seek comfort in a hot springs hotel in nearby Truth or Consequences. Once actually called Hot Springs, this interstate town changed its name in 1950 after the TV host of the popular *Truth or Consequences* game show announced he would air the program from the first town that agreed to the gimmick. Sixty years later, the economy here is still somewhat grim and still mainly hotels, from cheap to expensive, specializing in hot springs and tubs. Our hotel's pool is right next to the Rio Grande so that we can jump from steaming water to the comparably cold river, careful to keep hold of the rope tied to a wooden dock. The channeled Rio Grande is deep here, bank to bank, and if you let go of the rope you'd be swept down to Caballo Lake State Park (or really just a few yards to beach on someone's private property.) We're holding the rope and laughing in Pig Latin. *Ontday etlay ogay. Urrentcay ongstray!*

September 6

I don't believe it. I call for Peter. Two circular holes in the terrarium with the cliff bank soil. They've been made with the smallest of drill bits. They are perfect. I have babies. I rush to the refrigerator for my chilled mini-mealworms. Now I need a razor to cut one of the thinnest mealworms into thirds. The adult tiger beetles don't try to escape when I lift the lid and use featherweight forceps to drop dismembered mealworm parts into the one millimeter holes. I imagine my Western red-bellied tiger beetle larvae inside, anchored in their tunnels, scything tiny mandibles. Manna. Momma. Food.

I take a photograph. I e-mail David and Barry. I e-mail a few friends. My larvae have hatched! Thirty-one days from when I first saw mating to laying to digging their burrow, which corresponds to the guidebook's nine to thirty-eight days for most species. Peter and I embrace. A new standard of happiness.

We all know the truth about happiness. Win a big prize and feel important and, yes, that's happiness. But the next moment you can be happy, too, with the brightly colored socks you just bought on sale. Look at that flash of lime-green at your feet. Aren't you bold and attractive? And you got such a deal. Life is good with your green socks and happy feet.

September 7

The third instar we dug up from the banks of the Rio Grande is dead in its plastic tube. But there are two more circular holes in one of the smaller terrariums, the one with upland soil from the base of grass clumps.

September 8

My tiger beetles and I are driving an hour south to Deming, a small border town where my twenty-seven-year-old daughter teaches third grade. Even a short road trip has potential, like Jack Kerouac's arrow "that could shoot all the way out," the possibility of reinvention with music and the desert flashing past. Earlier this morning I caught and transferred the adult beetles from the two terrariums with larvae into my second large terrarium with sandy soil. I'm assuming these females have finished laying all the eggs they will lay. All my adults are likely near the end of their natural lives, in captivity for their last month, after being in the wild for an unknown number of weeks.

During the transfer, two beetles escaped into my office. I'll get them later, I think, because schoolteachers live by the clock and I'm presenting my citizen science project promptly at 2:30 p.m. For now I'm on schedule, listening to rock and roll from the eighties. Bruce Springsteen's *Born in the U.S.A.* Talking Heads' *Stop Making Sense.* And the Chihuahuan Desert arrowing past, yuccas like exclamation points rising out of grassland, grassland grazed down to soil, soil turned to clouds of dust. This is retro, not reinvention, and morally ambiguous; like other American traditions, the road trip has turned out to be a bad thing, using up gas and emitting carbon dioxide. I sing to my beetles about life here in the United States. Most of these songs, I tell them, are ironic.

My daughter is in her second year of teaching children in a public school and, as with other new teachers, the job consumes her. This is all she does. "Be like Coyote," she tells her class as they rush about putting away their last assignment. The seven- and eight-year-olds raise a forefinger and little finger (some European sign for cuckoldry?), which I learn later means "I am quiet now. I am ready to listen." And they do listen–very nicely, I think–to my speech about tiger beetles. Then they look at books on beetles as they take turns crowding close to the live beetles in the terrarium. They covet the close-focusing binoculars. They ask why the beetles are on top of each other. We marvel together at the big mouthparts. Before I leave, a girl presses a note in my hand.

> Since
> Is fun to do. Since is amracle. It's a ream. We love since. It's
> beatoful since.
> We love since!

I read this a few times before understanding that "since" is science.

September 14

I have three open larval burrows in the small terrarium with soil from the grasslands above Bill Evans Lake and ten open larval burrows in the large terrarium with soil from the cliff bank right by the lake. The grassland soil has fewer rocks and feels "softer" than that from the cliff bank, but both would be classified as loamy, an equal mix of sand, clay, and organic matter. The pH of these soils is 8 or slightly alkaline. The pH of two of the rejected terrariums, also with grassland soil, is 7 to 7.5, or neutral. The pH of another rejected terrarium is 6 or slightly acidic. The large terrarium with very sandy soil was rejected, as was the terrarium with very rocky soil. My hypothesis, then, has been partially confirmed; these tiger beetles probably don't lay eggs on soil near the water's edge, which is prone to flooding. Instead they have oviposited in the somewhat rocky, slightly alkaline soil of the cliff bank (which I had thought they would also reject) and less rocky and slightly alkaline soil from the hillside above.

In the last week, my first stage or first instar holes have opened and closed and opened again as the larvae plug their tunnels while digesting their meals. Brand-new holes have also appeared and then closed. I promise myself that next field season I will make a chart of all this.

Sometimes the larvae flip out parts of a mealworm they don't want. Sometimes I can see their heads flush to the ground, waiting for me to feed them. In other labs, captive tiger beetle larvae have been known to cache prey, taking down the food, and then returning in a short time to receive more. One researcher reported that a Williston's tiger beetle grub dragged twenty-one small adult moths down into its burrow within one hour and later threw out thirteen of them, uneaten. A second larva took down thirty moths and threw out fifteen in three days, with more moths found uneaten in the burrow after the larva had died. "Such behavior," said the scientist, "is probably not normal."

I am hoping that soon my holes will close and not open again until the first instar has shed its old skin and molted into a second instar, enlarging its burrow and entrance. Eating and growing from first instar to second instar can take as little as a week in some tiger beetle species, with the molting process requiring another week. I'm cutting up mealworms every day and dropping them into holes and no longer feeling, much, like a serial killer, reminding myself that insects are naturally born to the macabre. Parasitic wasps are forever laying eggs in the bodies of hosts, which the hatched larvae devour alive. Female spiders and praying mantis are forever eating their mates even as they mate with them. As the writer Annie Dillard once observed, "Fish gotta swim. Birds gotta fly. And insects gotta do one horrible thing after another." I'm not behaving so well myself, angling the head of the tri-dissected but still moving mealworm into a burrow so the worm seems to slither down of its own accord. It's just beatoful since.

September 15

At our second meeting of the How to Become a Leading World Authority Club, Allison explains how she has tagged eleven plants on her seventeen acres of land: two blue grama grasses, two side-oats grama grasses, one ponderosa pine, two piñon pines, two gray oaks, and two cottonwoods. She's varied their location, one in sun and one in shade, and through the coming fall and winter she will be looking twice a week at when seeds drop from each grass head and when needles stop forming on each conifer and when leaves fall from each deciduous tree. She inputs her data into the website for Nature's Notebook and sometimes looks to see what that means nationwide: what other grasses and trees are doing on the planet this fall.

Side-oats grama (photo by Elroy Limmer)

Ed has contacted the founder of an organization dedicated to processing edible green leaves and unlocking their nutrients to supplement the diet of poor families in poor countries. Ed's new mentor remembers standing in a yard in rural Nicaragua talking to a malnourished woman at the foot of a moringa tree rich in protein, iron, and vitamin A. If that woman had only known to grind up a mix of moringa leaves! No, this man wrote back to Ed, there hasn't been any work with growing leaves hydroponically.

I talk about the generosity of scientists who so often respond this quickly, happy to share their work and see the connections build, happy simply that you've read their paper. I imagine all those connections made visible: the study of a moringa tree linked to hungry people in Brazil linked to swallows who travel from South America to my front porch linked to children slipping notes in my hand, a crisscross of overlapping threads and all the threads glowing so that the planet becomes a gossamer veil of light or perhaps a solid brilliant glowing ball. I don't actually say any of this out loud.

Western red-bellied tiger beetle larva

September 17

Spoonful after spoonful, the dirt from my terrarium spills onto the newspapers laid on the table. I'm looking for a glimmer of movement, something very small and pale, with the same general features shared by all tiger beetle larvae: a relatively large armored head with scissoring mouthparts and a white body kinked where the back hooks grow. No human being in the universe has ever seen (or at least documented) this creature before. It's possible the larvae of the Western red-bellied tiger beetle actually have purple fur or whirligig eyes or the ability to communicate telepathically. This is something I'll know in just a few minutes.

I use a paintbrush to spread out the dirt.

Nothing. The kitchen table is a lunar landscape. Getting out the close-focusing binoculars, I walk past craters and rock. A bleak, lifeless world. And so lonely. One carbon-based creature calls out for another.

Then the heart stutters, and I feel like a UFO-er seeing light on the horizon. The first instar of a Western red-bellied tiger beetle shakes off the trauma of being upended from its burrow and starts to crawl, searching for soil where it can dig a new tunnel. It may look like every other tiger beetle larvae, but I sense something special about this fifth-of-an-inch-long (five millimeters), purposeful animal heading for the edge of the newspaper. Now another first instar starts to move as well. *I am the first person on the planet to ever see this,* but I can't let that go to my head or my larvae will fall off the top of their world onto the floor.

* * *

An hour later, and I have transferred four first instars into individual larval plastic tubes of dirt, where they immediately start digging their burrows. One other larva is circling the confines of a glass vial. I'm driving them all over to my friend Dick's house here in the Gila Valley. Dick is nearing seventy and is the full-time caretaker of his disabled wife, who has been declining into a near-vegetative state. Both his computer and photographic equipment are old and jerry-rigged. Still, there is no one else I know who could photograph an animal half the size of my fingernail. We spend the rest of the morning coaching the larva, which keeps having to be gassed with carbon dioxide to keep it still. Modeling is a high-risk job. We put the now-dead grub back in its vial, with alcohol, to keep it preserved as a specimen. I am hoping that these photographs will be good enough to reveal important taxonomic details like the shape and relative size of the back hooks, the size and number and placement of eyes, and the relative lengths of the short antennae segments. I leave behind two more larvae in their plastic rearing tubes.

As we walk out to my car, Dick and I talk about his dream of photographing the wildlife of Yellowstone National Park, surrounded by an abundance of life: herds of bison and elk, packs of wolves, convocations of eagles. I have been urging him to send David Pearson the photo of an unidentified tiger beetle that Dick found in his backyard, and I ask about that, and Dick smiles. David has confirmed the insect to be the pygmy tiger beetle, one of the smallest tiger beetles anywhere, about three-tenths of an inch long.

I exclaim, "The pygmy tiger beetle previously thought to exist only in Mexico and southern Arizona?"

"No longer," Dick crows. "We now know that the pygmy tiger beetle also lives here in the Gila Valley of New Mexico."

September 26

Two second instar holes appear in the big terrarium. My first instar larvae are molting into the next phase! These circles are noticeably larger, and I start cutting the mealworms in half rather than thirds.

September 30

Another second instar hole. This really has been an exciting month.

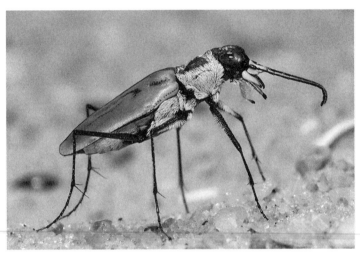

Eastern beach tiger beetle (photo by Kevin Fielding)

October 2011

October 2

Peter and I are a week in Georgia, where I am giving a nature-writing talk for a group dedicated to preserving the native plants of Georgia's coastline. The founder of the group lives on a barrier island that is also a private wildlife refuge: forests of live oak and palmetto, salt marshes with roseate spoonbills, sweeps of sand and rock and ocean and wind, and the cream-colored Eastern beach tiger beetle. Only recently I saw a YouTube video about this common beetle in which Professor Alan Harvey explains how one of his graduate students came back from her spring vacation on Cumberland Island National Seashore chatting about all the "neat things" she had seen–the wild horses, the tame armadillos, the tiger beetle larvae rolling down the beach. "Wait," Professor Harvey said. "Go back to that rolling tiger beetle larvae." No one knew then, although we all know now, that the larvae of the Eastern beach tiger beetle can leap into the air, form several tumbling somersaults, and land in a hoop that careens down the sand at high speeds for long distances. No entomologist had suspected that any insect did anything like this.

The title of my talk is "The Physics of Beauty," taken from the seminal *Sand County Almanac* by conservationist Aldo Leopold:

> The physics of beauty is one department of natural science still in the dark ages. Not even the manipulators of bent space have tried to solve its equations. Everyone knows that the autumn landscape in the north woods is the land, plus a red maple, plus a ruffed grouse. In terms of conventional physics, the grouse represents only a millionth of either the mass or the energy of an acre. Yet subtract the grouse and the whole thing is dead.

During my stay on St. Catherine's Island, I don't see any larval hoops, only the beautiful Eastern beach tiger beetle adult with its hieroglyphic pattern of three wavy brown lines. At the end of the day, these insects are as memorable as the alligator I also saw by the side of the road, and that's an interesting insight into the physics of beauty. The alligator was a massive predator whose mouth opened wide as if to embrace some stray part of my body, a hand or foot. If I had gotten closer, out of the car, the alligator might have prompted some memorable adrenaline, his ability to go from sluggish to lunge, that great muscular tail and smell of algae. In these blue-green marshes numinous under a yellow sky, the alligator may well be someone else's grouse–but he is not mine. He does not belong to me like the clotted cream-colored Eastern beach tiger beetle.

It's an interesting insight into the hippocampus, that part of the brain where learning starts and memories form in a self-organizing system that is flexible, overlapping, and redundant, with ten billion nerve cells linked by branching dendrites allowing for a thousand trillion connections. Using the brain changes the brain, strengthening some of these connections and not others. Now the Eastern beach tiger beetle fits neatly into place in this new land of tiger beetles rising up in my prefrontal cortex, this sprawling construction filling the skull's living room floor. The physics of beauty is really the biology of beauty, and the biology of beauty is what we claim as our own, what we build *inside our bodies* to resonate with what we see outside in the world. Now I understand that almost everywhere I go, for the rest of my life, I will see tiger beetles. Everywhere I go, because of that, the world will be more beautiful.

October 11

From Savannah, Peter and I fly to Washington, DC, to celebrate my father-in-law's ninety-second birthday. I also have an appointment to see the tiger beetle collection at the National Museum of Natural History, where I'm guided by entomologist Terry Erwin, a man famous for a 1982 study in which he suggested as a testable hypothesis that there were not several million insect species on the planet, as then believed, but several tens of millions.

In a "kill 'em and count 'em" experiment, Terry dusted the canopy of tropical rainforest trees with pesticide and sampled the insects that fell below. From a single tree species, he found over 1,000 species of beetles and estimated that 160 of these were specialized to the canopy of that tree. Beetles represent two-fifths of insect species, so 400 other insects were also likely specialized to the canopy, with another third of insects below the treetops–altogether, 600 insect species endemic to a single tree species. If some 50,000 tree species exist in a tropical forest, that would mean some 30 million species of insects.

This number depends on a few assumptions, and many scientists today put the world's estimated number of insect species as closer to ten million. Since we have described only 1 million insects so far, that leaves 9 million to discover. In the United States, we have described 91,000 insect species with an estimated 73,000 to find. (In 2009, entomologist Jeffrey Lockwood spent thirty-six hours sampling insects in the Red Desert of Wyoming, a fact that immediately made him the world's leading authority on the region's arthropods. He guessed that the Red Desert contains some 5,000 insect species, dozens of which were previously unknown.)

Of beetles, we can point to 350,000 species worldwide, with over 2 million to go.

The task seems biblical.

The National Museum of Natural History's collection of beetles is one of the world's largest, with more than 7 million specimens. The tiger beetles are hidden in gray metal drawers in gray metal cabinets that present a solid wall with a single narrow corridor and an automated system that shuts one corridor to open another, one passageway at a time as the walls close together. Terry Erwin shows me how to operate the automated door of the "compactor," the opening and closing of the single passageway, and the numbering system for drawers–organized by genus–that can be slid from each tall metal cabinet. When I ask him if tiger beetles are among his favorite insects, he says, "Not at all. They like sun, and they run fast, and I don't." When I ask him about the role of citizen scientists, he says that citizen scientists helped him write his latest *Tiger Beetles of the Western Hemisphere,* co-authored with David Pearson, and he gets pretty lively showing me photos in the book that were sent to him by amateurs. Where would he be without this kind of help? Why else would he be helping me today?

Terry agrees with David Pearson that tiger beetle populations are important for a number of reasons. People have been collecting tiger beetles since the nineteenth century, and their popularity has made them one of the few insect groups for which we have long-term, worldwide information. Such large collections represent a predisturbance baseline by which we can measure change. In Europe, for example, records accumulated over the last 150 years have already pointed to habitat decline we would not have noticed otherwise. Moreover, as bio-indicators or markers of diversity, tiger beetles are relatively easy to census during the season of adult activity. When scientists wanted to assess the status of a protected area in Peru, ornithologists took almost five years to document 90 percent of bird species; David Pearson and his crew found the same percentage of tiger beetle species in the first fifty-four hours.

Finally Terry waves me off toward those gray metal cabinets, "Have a look then. I'll be in my office."

Have a look then. At last I'll see the Rio Grande tiger beetle, not alive but nicely pinned. I can indulge in flash: the patterned orange, turquoise, and purple of the giant riverine tiger beetle of Northern India, the metallic green head of the Bolivian ornate tiger beetle, the bold colors of the tellingly named festive tiger beetle, splendid tiger beetle, and beautiful tiger beetle. Later I'll search out the world's biggest tiger beetle from the genus *Manticora*. And the hairiest, the hairy-necked tiger beetle that lives in New Mexico. I wonder how long Terry will be in his office. Three or four days? Maybe he'll let me camp out in these shifting corridors, one corridor at a time, set up a little cookstove and make cowboy coffee.

October 14

The day before we go back to New Mexico, I get two e-mails. One is from Dick. After many years of being an invalid, his wife has died. I am surprised. Maybe Dick is surprised. He has been married to this woman for forty years. Her life has been his life, especially in this last decade of caretaking.

And my friend Sarah writes me about my larvae. Sarah has been babysitting. For a week all the holes were closed and then, after I advised heavier watering, they opened again. Sarah is relieved.

From: Sarah
Sent: Fri, 14 Oct 2011 8:58:57 PM
Subject: Babes Rebound!

I fed SIX larvae today–and got to see four of them! Two of the tubes have tunnels, as I mentioned earlier, and though those larvae didn't make an appearance, they made their worm pieces disappear. In the terrarium, I misted much more heavily this morning and it seems that the larvae responded to that. A couple of hours ago I took a look inside the terrarium and saw a hole. And then noticed another one. And another! And then saw that a much tinier dot was indeed a hole. I got to see a glimmering black head inside each of these; in fact, the head filled the tunnel entrance at times. It's much easier to feed worm pieces into these holes rather than the larval tubes, and I got to see the worm start shaking and moving and get yanked down inside. The hole I spotted first, which seems bigger than the others, contains a larva that did two things: (1) it would take the worm

Third instar tiger beetle burrow holes with two larvae waiting for prey (photo by Sharman Apt Russell)

down and then a few minutes later would eject it (this I didn't get to see–but it happened three times), and (2) it was doing some excavating. I was able to witness it depositing some sand outside its hole.

Sharman, this was an absolute thrill. What a privilege to get a glimpse into this world of theirs. Kevin's joking that you're going to have to fight me to get these guys back. But that's only a joke.

October 16

The much tinier dot that Sarah saw is a new first instar burrow, perhaps from an egg laid just before I emptied the terrarium of adults on September 13, or from an egg that took longer to hatch for some reason. There are also five open second-stage, or second instar, holes. I take out one second instar for Dick to photograph. This is a repeat from a month ago. *I am the first person in the universe to ever see the second instar of the Western red-bellied tiger beetle.*

Dick says he welcomes the distraction. And he is distracted, dropping the larva on the ground–"Oops," we say together–repeating himself, forgetting at times what lens to use. I was only gone two weeks. For Dick, everything has changed.

October 18

This week a young couple in the Gila Valley opens the door of their unroofed shed to find all their goats, two nannies and two kids, slaughtered by a mountain lion. That is sad and alarming, especially since this family has young children who play in the yard and walk freely to the shed. The livestock deaths are reported to the state game and fish department, and now the law is inexorable–the animal responsible must be found and destroyed. Lions cannot be relocated because they simply come back, or cause new problems elsewhere, or suffer and die in the unknown habitat. Duly, two game and fish officers set out another live goat as bait, but the lion kills that goat too and escapes the trap.

Goats are popular livestock here among an alternative community trying to raise their own food. On the road where we live, immediate neighbors keep goats to the north and south, and I remember fondly the days when I also had goats and an alternative lifestyle. That was in the 1980s, when Peter and I were back-to-the-landers, believing we were on the cutting edge of social change. Children of the city and suburb, we were part of a larger cultural conversation, wanting to root our lives in soil and sun, to make the world better by making our personal connections to the world more direct, in the shape of an onion or adobe brick. It was in the air. Earth Day. Deep ecology. Bioregionalism. So we built our house of mud and had a too-big garden, too many goats, two homebirths– a daughter and a son–and too much goat cheese in the refrigerator. Our illusion that we could live off the land lasted a few months, or maybe a few years, and then Peter began a series of jobs, and I started teaching at the small university in Silver City, something I still do today. Eventually we sold the original homestead and moved thirty miles west to Silver City while our children finished high school. Then we moved back to the country, the Gila Valley, thirty miles farther west.

Meanwhile the larger cultural conversation shifted to green cities, the hope now for our relationship to the planet. Green cities are where 85 percent of Americans live, where humans will use the least resources and emit the fewest greenhouse gases, where creativity sparks in the diminished spaces between us, where we'll contain the damage of overpopulation. The big changes that need to take place are in the meta-systems of politics and law and business, and my individual connection to the land is less important than I once thought. The role of the rural West is less important than I once thought, and living in the rural West is less ecologically sound than living in places like New York City or Portland. I am not unhappy that the ideas of my youth–the very arc of my life–have been proven wrong. I'm only relieved that the cultural conversation is still alive. I'm pleased hope still exists.

And I believe those of us with roots in the country do have continued value as stewards on the ground. We are caretakers of landscapes relatively wild and free of human dominance. Part of the mosaic of human diversity, we are people who still have to think about mountain lions.

For that reason, I'm glad for a new citizen science program in my area called Wildlife Linkages, organized by the conservation group Sky Island Alliance, whose goal is to document major predators and endangered species in the Southwest. In Arizona, the accumulated data of their volunteer trackers has already been used to identify wildlife corridors that can be integrated into human development. Increasingly these corridors are included in county and state plans, and in highway construction and border projects–working toward a future where animals can move safely through urban and rural communities. In New Mexico, Sky Island Alliance is advertising for trackers who will walk the Burro Mountains, almost midway between Silver City and my home in the Gila Valley. (Yes, I think, that *would* be fun. That would be intriguing.)

October 20

I need a good specimen of the second instar of the Western red-bellied tiger beetle. The first instar that Dick and I killed while photographing was preserved in alcohol only, an inferior method since the alcohol will eventually degrade the body. The method preferred by David Pearson and Barry Knisley is to drop a live larva in a solution of 10 percent formalin just brought to boil and removed from heat. After thirty minutes, you take out the instar–which died instantly–and soak it in water for five hours. Then store your specimen with data labels in vials of 70 percent alcohol.

Formalin is 40 percent formaldehyde, with highly toxic fumes. My usual supply companies are happy to mail digital calibrators and terrariums but won't send formalin unless I provide proof that I work for a lab. In the end, I go over to Western New Mexico University, where I teach creative writing, so that a biologist can set me up with both formalin and the proper ventilating equipment. I'm shocked when my eyes burn and tear simply transferring the instar from its 10 percent solution to soak in water. As Barry Knisley has warned, "This is nasty stuff."

But I'm immeasurably pleased at my vial of a plump, floating second instar with the data strip curled inside, the name of species and date and place collected written on special paper with a special pen. Prompted by an e-mail from Barry, I'm rethinking again my contribution as a citizen scientist.

From: C. (Barry) Knisley
Sent: Mon, 26 Sep 2011 1:46:47 PM
Subject: RE: First instars
Sharman: Thanks for the photos, which look excellent. Not sure how Dave and I may have overlooked what is known/not known about your Western red-bellied, C. sedecimpunctata. In our 1984 monograph on larval tiger beetles of southeast Arizona, we include formal descriptions of most larvae, including one third instar of sedecimpunctata. Unlike most other species, we did not have first and second instars to describe and only one third. My rearing of this species was not very successful, producing only the one third instar, which I described. I am not sure how this relates to your planned work, but certainly descriptions of first and seconds and a redescription of third instars with variation among instars is warranted. I can send you a reprint of the paper if you do not have it. Also, of great interest is that we never found larvae of this species in the field, which was a great puzzle given how common they were. Barry

I will not, then, be the first person in the world to see the third instar of the Western red-bellied tiger beetle. But I do bear responsibility for describing the physical characteristics of first and second instars.

Also, there is the great puzzle of never finding those larvae in the field. We know these beetles disperse after the rains and can fly for some distance. The Gila River, its banks swarming with Western red-bellied tiger beetles, runs through three million acres of national forest land. If I commit to searching a million acres a year, I could do this as a three-year study.

October 30

Learning about tiger beetles is a physical process that takes place in the brain—in my case, a middle-aged brain. Such a brain has more trouble than before remembering specific words and focusing on new information. It's losing speed and mass, shrinking by about two percent every decade. At the same time, the middle-aged brain is better at certain tasks, such as seeing the big picture and making connections based on a breadth of experience and knowledge. Middle-aged brains vary enormously, but a good percentage of us compensate for the deterioration of aging by using both sides of our frontal lobes to solve a problem instead of just one lobe at a time. Researchers actually see this in brain scans. Young brains athletically use only the left side to encode and learn a new word and then only the right to retrieve it. Middle-aged brains also use the left lobe to encode and learn a new word and then *both* lobes for the harder job of retrieval. All this might account for the results of a study that tracked six thousand people for over forty years: in four out of six categories tested–vocabulary, verbal memory, spatial orientation, and inductive reasoning–most people did their best when they were between the ages of forty and sixty-five. We're doggedly using two hands instead of one.

In our efforts to absorb new material, we also gravitate naturally to what comes easily. We learn, after all, by building on what we already know in an emotional context in which the subject interests us. I've spent a lifetime immersed in story–books, TV, movies–and reading and learning stories about beetles is comfortably familiar.

For example, the idea of medieval trials against insects amuses me. In 1320, in Avignon, France, the Church had proceedings against the larvae of cockchafers, or melolonthine scarabs, which were damaging food crops. Before the trial, priests visited the area to summon the larvae to appear before the Bishop on pain of excommunication, advising the grubs of their right to counsel. Meanwhile, an advocate was designated whose defense of his clients–when they failed to appear–was that as creatures of God they had a right to eat. Moreover, their absence at the trial was due to their not being guaranteed safe passage. The judges disagreed and resolved that the larvae not only had to quit ravaging crops but leave the farming area entirely. Larvae who failed to comply would be killed. (In another medieval trial, offending larvae were excommunicated first.)

This kind of fairytale information fits nicely into parts of my brain.

I also understand how people can wear beetles–drawn to the iridescence created by the microsculpture of the insect's exterior shell, the effect of light bouncing around minute spaces. The shawls of tribal Thai girls glisten with green beetle wings. In Sri Lanka and Mexico, women pin live beetles to their clothing with a small chain around the insect's leg. In the nineteenth century, trade documents from northern India list millions of coppery beetle parts sold for use in textiles and ornaments. In the late Paleolithic, twenty thousand years ago, pendants were made in the shape of a family of beetles known for its brightly colored, jewel-like species.

Certain aspects of beetle behavior make sense to me—something I would do if I were a beetle. Might I not create an umbrella of leaves and feces and carry it over my head as a screen against predators, like the larvae of the tortoise beetle? Or light up like the female firefly (which is really a beetle) in order to signal males of my species that I am waiting, waiting, waiting—and then glow seductively to attract males of other species so that I can eat them?

When attacked or frightened, I think I might equally feign death as so many ground beetles do, lying upside down for long minutes with legs and antennae pathetically curled under. This works especially well for predators who prefer living prey; in one experiment, spiders ate death-feigning beetles only 38 percent of the time and not at all if obviously live beetles were available. Other predators will eat anything, of course, and so some click beetles have a back-up plan. A "dead" click beetle about to be picked up by a bird will arch its head to force a throat projection into a groove on its underside. Then the beetle straightens its body, unsnapping the connection, the momentum of which produces a loud click and causes the beetle to jump as high as six inches. Startled, the bird flaps away. Alternately, a short-winged mold beetle enters an ant's nest as an edible cadaver only to miraculously come alive, tricking the ants into caring for it even as it consumes their eggs and larvae.

Medieval courts and dancing girls. Rapunzel, Bluebeard, Odysseus, and the Trojan horse. Remembering these stories is literally child's play—even if I forget the actual name of the beetle, its Latinate species and family.

But experts say that the middle-aged brain shouldn't be playing or learning the easy way. We need to keep growing brain cells by challenging ourselves to get off the beaten neural pathway. As one scientist said, "Crack the cognitive egg and scramble it up." Another researcher adds that adult learning should include a "disorienting dilemma" or something that "helps you critically reflect on the assumptions you've acquired."

Today as I read the 1984 monograph Barry promptly sent off in the mail, "The Biosystematics of Larval Tiger Beetles of the Sulfur Springs Valley,

Arizona: Descriptions of Species and a Review of Larval Characteristics for *Cicindela (Coleoptera: Cicindelidae)*," I feel a disorienting dilemma coming on. Describing the physical characteristics of tiger beetle larvae targets skills and abilities I've never had, such as learning a new language, obsessive attention to detail, and good eyesight. I read and reread the section for the Western red-bellied tiger beetle.

> *Cicindela sedecimpunctata* Klug
> Measurements: TL 14.0; W3 1. PNW 2.3; PNL 1.5; FW 1.2; FL 1.2
> Color: head and labrum black; pronotal disc and cephalolateral angles black or black-brown. Antennae dark brown, maxillae yellow brown to brown; labium yellow-brown basally, brown apically. Mandibles yellow-brown basally, apices black. Mesonotum and metanotum light brown posteriorly. Dorsal cephalic and pronotal setae white to light brown; other setae yellow-brown.
> Head: Dorsal setae prominent; U-shaped ridge on frons with 2 setae. Antennal segment 1 with 5 setae; segment 2 with 9 setae.
> Pronotum: Pronotal setae prominent, 13 pairs; 9 pairs long or distinct, others small or minute (Fig. 116). 14-15 pairs of cephalomarginal setae.
> Abdomen: Scletorized setae prominent. Third tergites with 16-19 setae (Fig 117). Median hooks with 3 setae; inner hooks with 2 apical setae, spine one third to nearly one-half total hook length (Fig 118). Fifth caudal tergites with 42 setae, 21 stouter; epipleura with 8 setae. Type IA hypopleuron. Ninth eusternum with 2 groups of 3 setae (may be 4 in 1 group) on caudal margin (Fig. 119); pygopod with 7-0 setae on each side.

My heart sinks and my middle-aged brain thinks, "I'd rather stab my cortex with a fork." I don't have the proper equipment or the emotional engagement. I'm having a good time walking up and down rivers and visiting third-grade classrooms and showing off the picture I carry in my wallet of my first instar. But counting forty-two setae on the fifth caudal tergite sounds as much fun as reading Proust backwards.

I pull out Dick's photos and David's books. The third tergite is an abdominal section two segments down from the easily seen back hooks. The eusternum is the second to last section of the larva's body, and the pygopod is the very tip. I could learn more from a detailed illustrated book like the massive two-volume *American Beetles*. I'll need a way to measure the insect using my neighbor's microscope and, yes, my supply company sells a miniature scaled ruler for that purpose, calibrated to 0.1 millimeter.

I'm certain I'll need both lobes to do this.

Fall scene on the Rio Grande (photo by Elroy Limmer)

November 2011

November 8

A third instar hole appears in the terrarium. The entrance to the burrow looks huge, about 2.5 millimeters.

November 12

I take out the third instar for a photo shoot. The larva is 10 millimeters, and Dick and I exclaim. What a monster.

November 18

A second third instar hole appears after I water the terrarium more heavily than usual.

November 26

Midmorning, and one of our neighbors, a rancher and friend, crosses the bridge over the irrigation ditch and walks toward our house carrying a big rake. Jerry is looking for the mountain lion his cousin just shot out of a cottonwood tree a quarter mile up the ditch. The animal dropped into the water and floated down and has snagged up somewhere around a branch or root.

Last evening this lion killed another neighbor's billy goat and dragged it across the road to Jerry's property. Jerry's cousin happened to be visiting and happened to have a mountain lion hunting license as well as hound dogs. Tracking and treeing the lion wasn't hard, since the female seemed to be living here in the cottonwoods and ash and walnut trees that line the ditch, in what I think of as my backyard. We assume this is the same lion Peter saw some months ago, the same lion who last month killed the young couple's nannies and kids, with Jerry confirming that in his experience mountain lions "do love their goat meat." Soon after we talk, Jerry finds the corpse, he and his cousin lay her out on the ground, and the hunter says admiringly that she is the biggest cat he has ever seen.

Rio Grande and Sharman Apt Russell (photo by Peter Russell)

A license to hunt mountain lions costs forty-three dollars for a New Mexican resident. Killing the animal usually requires skill and a pack of trained dogs, first for tracking and then for treeing, a situation in which the lion often settles comfortably in the crook of a branch and stares at the excited barking below, not understanding what's coming next. We seem to be seeing more and more lions in the Gila Valley, perhaps because many of our old-school hunters have retired or lost interest or perhaps for some other reason. Most of these mountain lions are simply passing through, up and down the river, with young males looking for new territory. But a few females also raise cubs here, in what they think of as their backyard.

Later in the day we learn that, in fact, this is a second goat-killing lion; the first is still at large. This billy goat, too, was relatively easy prey, in an unroofed pen with a fence that my neighbors thought was high enough. I'm not happy about this lion's death but have to remind myself: mountain lions are not endangered and not threatened, and this one may well have become too habituated to living among humans, lounging in the cotton-woods along the irrigation ditch, twitching at the smell of that billy goat so close. I'm glad, at least, that Jerry also didn't look happy as he searched for the big drowned predator. He didn't have the gleam of the victor. This was just something that had to be done.

Gila River (photo by Dennis Weller)

Winter 2012

Throughout the fall, my tiger beetle larvae have plugged their holes and stopped eating, usually while molting into the next stage or instar, from first to second and second to third, but sometimes also in response to an environmental condition, such as when I water the terrarium. That flooding may have prompted a quiescence or suppressed metabolism. "Let's wait," they might have thought if they could think, "until the heavy rains have stopped."

Quiescence is reversible and does not require much preparation or physical change. Diapause—sometimes called hibernation or dormancy—is different. Preparations for diapause begin early, prompted by hormones. Buzzers sound. Alarms ring. Biochemical switches turn on and that turns on more switches. Although we don't know much about diapause in tiger beetles, we can look at what other insects do. In the larva of a butterfly preparing for winter, the gut is evacuated, blood thickens with an anti-freezing agent that acts like glycerol, water content in the body decreases, and free water converts to a gelatinous colloid. In temperate climates, diapause is usually signaled by changes in day length; the pigments in a caterpillar's blood are sensitive to light energy and help count time. The larva or caterpillar of the cabbage white butterfly, for example, can distinguish between fourteen-and-a-half and fifteen hours of light. Forewarned, the caterpillar seeks out a protected site where it remains safely immobile. Since tiger beetle larvae are overwintering in underground tunnels, the cues of a shortening day may be replaced by other factors, such as lower temperatures. Once diapause is triggered, there's no turning back until the insect is cued again by the right signal of a longer or hotter day.

In North America, tiger beetles in the genus *Cicindela* are classified into summer-active species and spring-fall active species. The Western red-bellied tiger beetle is a summer-active species. The beetles overwinter or diapause as larvae. The larvae eat and grow and pupate in the spring and emerge as adults in the summer when they mate and lay eggs. These eggs hatch and become larvae that will now overwinter. A summer-active species might take two or three winters before its larvae get enough food to go from first instar to second instar to third instar to pupation.

I have two third instar larvae in my terrarium that have been quiescent off and on throughout the fall like children grown too silent in their room, until I wonder what they are really doing. These larvae have not had the wild life of most tiger beetle larvae. They live in a terrarium in a house that does not freeze at night. They have had no cues of decreasing daylight or colder temperatures. They've been well fed and are molting steadily. This winter I would like them to skip diapause and form their pupae, which I would then like to dig up and photograph.

What else do I have to do now that tiger beetles have disappeared from my world?

Barry Knisley says, "Sure, see if your larvae pupate." He is on his way from Virginia to Naples, Florida, working on the newly discovered Florida tiger beetle, then to the Galapagos Islands. David Pearson is going to Bolivia. That's another option for entomologists when their insects are in diapause: go somewhere they aren't.

✳ ✳ ✳

When the weather turns cold, citizen scientists can travel like Barry and David or catch up on their reading or join a citizen science project that embraces winter–watching ice in a nearby lake or monitoring snowfall. IceWatch USA is part of the National Phenology Network and these volunteers, like Allison Boyd looking at trees in her backyard, help scientists document climate change. Capitalizing on social media, the SnowTweets Project asks people to measure snow depth in their area and tweet the data to the web; the information is picked up and mapped in real time and helps calibrate other snow-measuring systems, such as satellite remote sensing and simulation models. Other projects also map winter precipitation, with the observations used to predict weather trends.

As a holiday greeting, the website SciStarter sends me an e-mail that starts "The Twelve Days of Christmas" going round in my head like a carousel. On the first day of Christmas, stem the threat of invasive pear trees in Missouri (*a paaartridge in a pear tree*); on the second day of Christmas, record the nests of turtle doves in the United Kingdom (*twooo turtle doves*); on the third day of Christmas, click the Greater Prairie Chicken Project (*threee French hens*); on the fourth day of Christmas, participate in Audubon's Christmas Bird Count, the world's longest running citizen scientist project (*fouur calling birds*); on the fifth day of Christmas, monitor the water quality of the Yuba River, affected by gold mining (*five go-oh-olden rings*); on the sixth day of Christmas, look at geese density in saltwater habitat (*six geese a-laying*); on the seventh day of Christmas, report sightings of black swans (*seven swans a-swimming*); on the eighth day of Christmas, go to the Milky Way Project (*eight maids a-milking*); on the ninth day of Christmas, watch thirteen hundred young ladies set a new Guinness World Record for the world's largest cheer for science (*nine ladies dancing*); on the tenth day of Christmas, protect frogs at the Association of Zoos and Aquariums' FrogWatch (*ten lords a-leaping*); on the eleventh day of Christmas, enjoy SciencePipes at the Cornell Lab of Ornithology, a free service that lets you connect to biodiversity data (*eleven pipers piping*); and on the twelfth day of Christmas, count ruffed grouse with the Ruffed Grouse Drumming Survey (*twelve drummers drumming*).

This is one of my favorite Christmas carols, with that satisfying loop back to the beginning (*paaartridge in a pear tree*). I phone the founder of SciStarter, Darlene Cavalier, to thank her for what's she done to the rest of my day. She's a former cheerleader for the Philadelphia 76ers basketball team and also founder of Science Cheerleader, a website and blog promoting some two hundred former and current professional sports cheerleaders who are also professional scientists: Talmesha cheers for the Washington Redskins and has just completed her PhD in cellular and molecular medicine from John Hopkins University; Kristy cheers for the Atlanta Falcons and is a research chemist for the Center for Disease Control; Maria cheered for the Houston Rockets for four years and now is an electrical engineer for NASA. Science Cheerleader performs at conferences and school events as a way to encourage girls—especially those three to four million young US cheerleaders—to enter into the fields of science and math. Without doubt, the women at Science Cheerleader also want an excuse to get into that costume again.

Darlene started the SciStarter website in 2009, promoting obscure and small-scale citizen science projects alongside those that are well known and well funded. Among her favorite successes is the rediscovery by a citizen scientist of the nine-spotted ladybug, New York's official state insect, which lawmakers were trying to replace since it was thought to have gone extinct. She's also enamored of Firefly Watch at the Boston Museum of Science, mapping firefly populations across New England. Over the phone, she says, "In the very process of signing up for this program, you start to piece together aspects of firefly natural history. You need to learn the difference between male and female fireflies. You need to learn about the effects of pesticides and herbicides on firefly larvae. You need to start thinking critically." Darlene has four children and wants them all to think critically. Her current favorite citizen science is the Microbomes Project, now soliciting twenty volunteers who are relocating within the area of Chicago. Households with uncaged pets are not eligible. The goal is to collect the millions of microscopic skin and bacterial cells that humans shed every day and then determine how quickly that unique microbiome is established in the new environment. The results will show how humans interact with the surfaces of their homes and the impact we have on existing microbes.

Darlene believes that citizen science will become increasingly important as a crisis response network. In 2010, after the Deepwater Horizon oil spill in the Gulf of Mexico, the citizen science program eBird began matching their baseline data with the new information coming in about birds. They also developed an application for the iPhone that allowed people on the Gulf Coat to easily report the status of beaches and wildlife. Similarly, a week after the March 2011 earthquake in Japan that caused major leaks in the Fukushima nuclear power plant, the grassroots organization SafeCast started building a radiation sensor network of mobile and handheld devices near the exclusion zone. Increasingly, the infrastructure of citizen science will allow for such quick responses to large-scale emergencies.

It does make me want to cheer, kick up my feet, and flip backwards—something I'd like to be able to do in any case. But for many projects, questions remain. Is this science, a process that results in valuable information, or is it PR, a way to rally the public behind science? Especially when projects involve children, is this science or science education? And when the project does result in science, are citizen scientists simply extended tools, collecting data and making measurements—or, more rarely, as in the case of the Microbiomes project, human subjects in an experiment? Or are they people learning real scientific skills? And which answer to which question is bad or good?

The Cornell Lab of Ornithology is emphatic: their team of citizen scientists produces real science. In 2004, their House Finch Disease Survey discovered and tracked an epidemic of house finch eye infections in the Northwest. Since 1987, over forty thousand people from the United States and Canada have participated in their Project FeederWatch, with 1.5 million checklists that help researchers identify species and monitor changes in abundance and distribution. Based on this data, a 2008 article in the scientific journal *Ecography* reported that an abundant population of introduced collared-doves was surprisingly correlated with an increased population of four native dove species. Another long-running study, Birds in Forested Landscapes, connects acid rain to the decline of wood thrushes. Decades of observations and tens of thousands of volunteers have contributed to over sixty scientific papers. As Cornell's brochure says (chirps), "Birds are everywhere. Researchers are not!"

Galaxies are everywhere, too, and researchers are not, and one of the most successful citizen science programs remains Galaxy Zoo, which uses crowdsourcing to classify the two to five hundred billion galaxies in the universe. Online communities around Galaxy Zoo have begun to self-organize, and a culture of zoo-ites now communicates around the world. Their work gets astronomers associated with Galaxy Zoo access to some of our largest and best telescopes.

If the use of human volunteers has become a new form of scientific instrument, like instruments, we need to be checked for accuracy. Galaxy Zoo has so many observations they can duplicate them many-fold to minimize error. Australia's ClimateWatch asks volunteers to monitor changes in plant and animal phenology and then uses an algorithm to look for unusual data points, which are investigated. Many projects manually check a percentage of their data as quality control.

For some citizen science projects, a lot of people does not just mean a lot of data, but a lot of creativity. The online game Foldit has hundreds of thousands of players working at folding or designing model proteins on their home computers. When researchers first asked these gamers to remodel a specific enzyme to increase its contact with reactants, they received 70,000 designs. When they wanted to know how to stabilize the amino-acid "loop" in the enzyme, they got 110,000 answers. Using the best designs, they synthesized test enzymes and came up with a winner. As a biophysicist with the project says, "I worked for two years to make these enzymes better, and I couldn't. Foldit players were able to make a large jump in structural space and I still don't fully understand how they did it."

Such leaps aren't just about quantity–the number of designs submitted–since a computer could generate even more of those. Rather, folding amino acids into the right shape seems to require a uniquely human touch.

In a more aggressive search for human creativity, businesses and organizations at the online marketplace InnoCentive offer cash rewards for the solutions to specific research problems–how to retool a piece of lab equipment or track the progression of Lou Gehrig's Disease. All solutions are welcome, and no one is checking your professional background.

Clearly, citizen science can be real science, and that's important to more people than scientists. Real science is needed to produce real social policy. The mass of data being gathered about climate change is helping inform American politics. Projects like Invaders of Texas and Nature's Notebook monitor invasive species, which leads to better regulation. Water sampling everywhere, from the Chesapeake Bay to the Colorado River to the Puget Sound, alerts agencies to newly polluted areas.

Programs at the Cornell Lab of Ornithology are often specifically designed to further the immediate health and protection of birds. A distribution map from the Golden-Winged Warbler Atlas Project will determine refuge areas for that species. Information from Birds in Forested Landscapes is used by the Forest Service to develop recreational sites. The Cornell Lab's motto, "birding that matters," has virtually replaced the image of the self-absorbed birder checking off her life list with that of the selfless environmental activist.

And that's the self-image of more and more citizen scientists.

A permutation of citizen science, called the "new wave" or extreme citizen science, wants to even more actively connect people, data, and policy. Extreme citizen science looks for research questions that come from the needs of communities themselves. They are especially interested in populations that are not educated and not middle-class—the traditional portrait of the citizen scientist—in places where environmental problems are urgent. In Deptford, England, community members used hand-held devices to monitor and map noise pollution from a nearby scrapyard—whose license was eventually revoked. In Tanzania and Uganda, the Jane Goodall Institute helps villagers gather information about biodiversity, forest health, and natural resource use planning. In what they call community-based conservation, they hope to preserve a life for wild chimpanzees through promoting sustainable livelihoods for humans. (In January, I e-mail the Institute for the third time. I don't expect to hear from Jane Goodall herself, just an assistant, someone working in the office, maybe a volunteer?)

Finally in this list of virtues—what makes the renaissance of citizen science so powerful—there is education, where my prejudice shows like an old-fashioned slip. An educated public has the opportunity to create better science, vote for better laws, promote better social policy, and live in a world made more tangibly marvelous. Our relationships expand to include galaxies and tiger beetles. We have new venues for creativity and happiness.

In my daughter's third-grade classroom, a scripted curriculum now gives her half an hour a week for science, something she loves to teach. Emboldened by my success with tiger beetles, I visit SciStarter to find a project we can squeeze into her limited time frame. The Mastodon Matrix program, run by the Museum of the Earth in Ithaca, New York, jumps right out. For $18, they will send us a kilogram of the fossil matrix surrounding an 11,500-year-old mastodon excavated at Hyde Park, New York, in the year 2000. The point is to sift through the dirt in order to reconstruct the environment the mastodon lived in. It's simple and brilliant. Who doesn't love mastodons?

The museum suggests four sessions. First students watch a DVD about the excavation, which opens with the dramatized scene of hunters and gatherers spearing a computer-generated mastodon lumbering onto an ice pond. The animal sinks and dies, preserved in peat. In the next hour, the children poke through handfuls of the excavated peat with toothpicks, sorting out what they find into jars labeled ROCK (shiny black chert sometimes used in tool-making and colored metamorphic specks carried to Hyde Park by glaciers), PLANT (bark, twigs, cones, fibers, seeds, charcoal, and leaves), ANIMAL (shells, hair, insect parts, ivory, and bone) or UNKNOWN (now the imagination can take flight–angel eyebrows? spaceship debris?) For the third session, the students examine everything more closely, size and weigh the material, and put it all in labeled, sealed baggies to send back to the museum. Finally they imagine the life of a mastodon in the Late Pleistocene and create a class project like a mural or play or newsletter.

Maria's class watches a third of the movie, with suitable exclamations. She has twenty-one students, and we divide them into three tables of six to eight children working in pairs. Peter has joined us so there will be one adult per group. We pass out paper plates and on each plate dump a handful of the gray fossil matrix, more like chalk than peat, with lumps of clay-like marl containing tiny snail and clamshells. I can also see loose twigs and fibers and include some of those per handful. One child starts to break open the lumps with toothpicks, while the other records the data.

Almost immediately they are calling out, "Over here!" "A twig!" "A shell!" "Miss Russell!" "Miss Sharman!" "Mr. Russell!" The room is an aviary of excited cries, and I am circling my table, round and round, from child to child. "Wow," I say. "Isn't that amazing?" And no one disagrees, not the little girl dressed like a pop singer nor the boy already and clearly disenchanted with school: this spiral a tenth of an inch across, this perfect whorled shell, is amazing.

"I found a bone!"

"Is this a hair?"

Snails and hair are transferred to the jar marked ANIMAL.

"I need more dirt!"

"This seed is green!"

Plant materials go in the jar marked PLANT.

The thrill of treasure hunting. I'm frankly exhilarated. The kids are happy, too, and they are almost careful. A boy teases out a single long twig. Another holds his breath while he excavates a leaf. Only one jar is overturned, all that data scattered on the floor. Only some of the kids drop their snails and can't find them. Clearly this project is about education rather than science: the third-graders of America are not miniature paleontologists.

Even so, I read later in a paper in *Paleontographica Americana* that the nearly all-volunteer project, which has sent samples to three thousand classrooms and whose main purpose is to improve public and student understanding of research, found seventeen taxa of seeds that were not reported in other smaller but professional studies. More hands at work, even small hands, meant more data. (At the same time, not just Maria's kids may have dropped their snail and clam shells, since "the total number of mollusks picked out," the author of the paper writes, "does not seem to reflect expected natural abundances.")

A week after that first session, I do the sorting, sizing, measuring, and bagging and take the results back to the classroom. We sit in a circle and talk. Some of the twigs crushed at one end and broken on the other may have been eaten by a mastodon and then passed through its intestines. Elephants are similar to mastodons but not descended from them, with their own branch on the family tree. Like elephant tusks, mastodon tusks were made of ivory, which humans once used to carve into tools and jewelry. And like the mastodon, elephants are in danger of becoming extinct.

Next the kids draw a mural about the Pleistocene on butcher paper. One girl can do a mastodon silhouette as if born to the task, making me rethink the gender of some of those cave painting artists. Another writes a story which I think remarkably good and which she gives me:

> Searching for long time ago stof and mastodons bones rough draft.
>
> We listend to a real story, wache a move, search for ones in the soil, and cleand the bones and the other stof it was the coolest Monday I ever had. We learnd scince, mastodons disappeared long time ago and there where plants like today. I like the diging, the movie, and what we did. I think we should watch the hole movie next time we watch the movie. In the future I want to do more digging, find more fossils like I did on Monday, and look for more stof under the water, soil, desert, and dert. We found

rocks, twigs (eatend a little bit), and animals shells and more stof. I found a fosil that look like a tooth of an exciting animal or a rock. We did not know what it was a tooth or a rock? I thought if it was a tooth? We did not know what it was? It was a strange thing I found. What do you think it was?

In their lifetimes, these children will see the loss of the elephant in the wild. Some of them will live to the end of the century, when scientists say the average surface of Earth could be 10 degrees Fahrenheit warmer, with the American Southwest warmer than that average. They will live through the problems we talk about now—water shortages, massive shifts in human populations, loss of agricultural land, catastrophic wildfires, resource wars. They will see landscapes change and towns die—quite likely, the border town where they live now. They will live through the solutions we dream about and the failures we worry about. Some will go on to become scientists. Some will work as teachers and bureaucrats. Some will become citizen scientists.

It's a fast growing field. David Pearson tells me that when he and Barry Knisley first published their field guide in 2006, there were only a hundred or so tiger beetle enthusiasts in North America, most of them amateurs. Now he can hardly keep up with the information he receives from thousands of tiger beetle fans, all those new distribution records and natural history observations. Tiger beetle mania! He expects to hear even more from places like China, India, and parts of South America as their economies grow and hobbyists have time to spend on the avocation of science.

As David writes in the publication *Wings*, the journal of the Xerces Society, "The future of insect conservation is more and more in the hands of these professional amateurs, whose contributions should help guide future policy decisions and budget planning by professional biologists, politicians, legislators, and policy makers. Their passion for tiger beetles illuminates the ways in which insects and their admirers can advance conservation policy everywhere in our threatened world."

We all want to be part of something larger. We want to be part of a family, a community, a cause. We want to be part of something meaningful. Studies show that long-term happiness depends on this engagement. I personally want to advance conservation policy. I want to do real science. I want to learn more science. Equally true, as Cyndi Lauper sang back in the 1980s, like most girls, I just wanna have fun.

In the entomology doldrums of winter, Barry Knisley sends me this e-mail.

> From: C. (Barry) Knisley
> Sat, Feb 18, 2012 03:27 PM
> Sharman: Your success in rearing the C. sedecimpunctata was a great accomplishment, especially getting the oviposition and rearing them through. For the paper we did on AZ larvae, I reared most of the larvae described in that paper from field collected females in small (5" x 8") plastic chambers, one or two pairs of beetles per chamber. After first instars appeared, I transferred them to the tubes. For most species these worked well, but as you know I had little success with C. sedecim, especially not being able to get oviposition, ultimately only one-third instar. C. marutha was also problematic but was able to collect many of these in the field. Ideally, we would like to have about five individuals for adequate taxonomic description. I would suggest preserving them according to standard method, boiling in formalin or water, then to alcohol and photographing and/or making the standard measurements and drawings.
>
> On another note, I was extremely interested about the pygmy tiger beetle found in your area. As Dave noted, this is a very valuable new record and to my knowledge the first for the state. In my NM report (not the Rio Grande report but the one for all of NM tiger beetles) for FWS I noted this species was one of several that might be found in the state. Do you have that report? In any event, I would like the specific locality and collection date, and would be interested in searching for it when I am out there.

Which brings me to plans for this summer. Have been coordinating with my former students (seventies biology from Franklin College who were on several of my desert biology trips) to establish a plan. The dates will be about July 20 to 27 and we would plan to spend a day or two in your country. They also want to learn more about tiger beetles and my work and yours, I am sure. I assume this period works for you. We would also like to take you up on the offer for your guesthouse for a few nights if available? Itinerary not yet set but we would probably hit a few other areas in NM for tiger beetles. My plans are to do a tiger beetles of NM, expanding my report to FWS. Barry

A great accomplishment. It's the nicest e-mail ever.

As long as I'm online, I go over to SciStarter to look at any new citizen science projects. The razzle-dazzle never stops. I can map the connection between cells in the human retina, listen to whale songs, lend my computer in the fight against malaria, and measure the weathering of marble in graveyards. Many of these citizen science projects contain the word watch–NestWatch, ClimateWatch, IceWatch, OspreyWatch, Frog-Watch, Redwood Watch, Firefly Watch, Mountain Watch, and Wildlife Watch.

In 1864, Thoreau wrote, "The woman who sits in her house and *sees* is a match for a stirring captain. Those still piercing eyes exercised faithfully with her talent will keep her even with Alexander and Shakespeare."

Some hundred years later, Annie Dillard concurred, "We teach our children one thing as we were taught–to wake up."

* * *

By January 3, my remaining third instar hole had been closed for ten days, and I assumed it had pupated. But digging up this area of the terrarium, I find instead a big third instar–twenty millimeters–and another second instar. I leave the second instar in the dirt and photograph the third instar. For the next four hours, that instar wanders about the terrarium unable or unwilling to dig a new tunnel. When I water the dirt, the larva responds quickly, excavating into mud.

In a few days, more holes appear, the second instar and third instar, but also two first-instar holes. I'm quite surprised. Eggs can wait this long to hatch? David e-mails that how long tiger beetle eggs are able to stay dormant is not known. I cut up mini-mealworms and feed these four larvae.

By January 17, the third instar hole closes again and by February 1, the second instar hole and two first instar holes are closed. I want to see a Western red-bellied tiger beetle pupa and I wait until February 24, when I dig up the third instar larva. No pupa, but the larva is still healthy and manages to tunnel another burrow, yet again.

Barry's e-mail nags at me. At least "five individuals are recommended for adequate taxonomic description" and I only have one preserved for each stage. I decide to dig up everyone and put them all in alcohol. On March 3, I empty the terrarium, cup by cup, onto the kitchen table and spoon and sift through the dirt. I find one shriveled body, probably the large third instar, and nothing else but the remains of mini-mealworms. The smaller larvae have decomposed and disappeared. This is disappointing but not unusual. The mortality rate of tiger beetle larvae and pupae is high in nature and in terrariums, too. Even absent any cold, and with plenty of food, winter can be brutal.

Mountain mahogany (photo by Elroy Limmer)

Spring 2012

In the mountains of southwestern New Mexico, spring is still too cold for adult tiger beetles to be active. Theoretically, of course, the larvae of the Western red-bellied tiger beetle are (somewhere) slowly waking from the sleep of winter and opening their burrow holes, from which they are lunging out and grabbing prey, eating and growing, getting ready to pupate and emerge as adults before the rains this summer.

Meanwhile, inspired by Allison Boyd and the How to Become a Leading World Authority Club, I start Nature's Notebook, the project sponsored by USA National Phenology, watching and recording the stages of plants and animals. Here in the Gila Valley, Peter and I own six acres of river bottom land, which can be divided into layers: a view of the Gila River marked by a line of trees in the distance; a large field that we have the right to irrigate but not the time; our house, barn, and garden; the irrigation ditch bowered by two rows of thickly growing cottonwood, elm, box elder, and walnut; and a flat area covered with shrubs and native trees like mesquite and desert willow between the irrigation ditch and a two-lane country road. It is this flat area, a scrubby buffer zone, where I will document species for Nature's Notebook.

I pick seven plants from the program's list of preferred species: box elder, soapweed yucca, four-wing saltbush, honey mesquite, desert willow, and the two flowering annuals of silverleaf nightshade and scarlet globemallow. The protocol is to tag individual specimens and check them at least once a week for young leaves, increasing leaf size, flower buds, open flowers, pollen release, fruit, and fruit drop. When I record my data online, I am further asked to estimate: "What percentage of the canopy has leaves (5–24%? 25–49? 50–74%?)?" And: "How many flower buds are present (11–100? 101–1,000? 1,000–10,000?)?" Some questions read like the dreaded word problem in a math class: "What percentage of all fresh flowers (buds plus unopened plus open) on the plant are open? For species in which individual flowers are clustered in flower heads, spikes, or catkins (inflorescences), estimate the percentage of all individual flowers that are open." I know that stopping to examine each plant will take some time and I have been advised by Allison to start with a small group. So I don't include any of the grasses or invasive species.

There is another list for preferred animal species, and I'm pleased to see the punctured tiger beetle. In magical thinking mode, putting an animal on my list will increase my chance of seeing that animal. So I commit to recording information for the American robin, American kestrel, bald eagle, black phoebe, black-chinned hummingbird, broad-tailed hummingbird, bumblebee, cabbage white butterfly, cliff swallow, great blue heron, house wren, monarch, mourning dove, mule deer, olive-sided flycatcher, punctured tiger beetle, raccoon, red admiral, and sandhill crane. I skip a few species on the preferred list like killdeer that are only near the river and bighorn sheep that are only near the cliffs near the river.

Then I head out into the wide world, crossing over the irrigation ditch on a small footbridge. Immediately I am struck by how little I know. The box elder has male and female plants, its winged samara, or seeds, staying on through winter. The four-wing saltbush also has two sexes, a male and female plant, although the yucca, mesquite, and desert willow do not. There are a number of shrubs I cannot identify, as well as a few trees growing by the ditch, and innumerable annuals just beginning to emerge. I look back at my house. I am surrounded by the towering trunks of cottonwoods and black walnut and ash and hackberry–mighty leaf-clothed giants I see every day–yet I hardly know how they reproduce or live.

Phenology is too dull a word for what is happening here. For how I must search along a stem for the smallest of new leaves, peer into the heart of a bud, and rub my fingers against a catkin, trying to release its pollen. This is a one-to-one relationship, me and this catkin, me and this tree, me and any tree nymph herein.

In one of my earliest visits, I see the Ouroboros, the mythical snake so large it wraps around the world eating its tail. I only see its midsection, five feet of muscle thick as my lower thigh stretching across a game trail. Its coloring blurs in the grass and distance, something like a diamondback rattlesnake, which can be this heavy-bodied, or possibly a bull snake. I have binoculars for distant birds but not the close-focusing binoculars I need right now. I wait. I wait. I wait. I wait. Then, wanting

Left: Box elder; right: Wolfberry (photos by Elroy Limmer)

a better look, I move forward and the snake flows over the ground in a flattened sine wave, flicking its rattle.

Next time I bring those binoculars, which also work well for birds in nearby trees. I see a bullfrog in the irrigation ditch and two spiny lizards in dispute–bobbing, defending–and a blue bug on the box elder. The insect looks like a turquoise bead, its color the exact shade believed in New Mexico to ward off witches. Later, the closest match I can find on the Internet is the long-legged fly, except that its range doesn't include the American West and its head doesn't look the same. In any case, I don't think fly at the time. I think: *mirabilis!*

The only animals on my list I see are the white-winged doves, whose lament is sweetly, deeply familiar. I check off, too, the cliff swallows darting in and out of their nests on my porch and the black-chinned hummingbird at the bird feeder. This isn't cheating. Nature's Notebook has a category for bird feeders.

I'm also alert for animals Nature's Notebook doesn't care about. I'm walking trails made by javelina, a pig-like hooved and tusked native species delightfully wedge-shaped, the adults weighing about forty pounds. Javelinas travel in family herds, a father, a mother with young, and adolescents. Female teenagers act as nannies. Male teenagers are skittish and sometimes run at an intruder, not bravely but because they panic and don't see too well. From the ditch on my right, I hear the bark of an Arizona gray

Fox (photo by Elroy Limmer)

squirrel. A few years ago, before the population was dramatically reduced by rabies, I might have seen a fox. Mule deer and skunk also walk these trails along the irrigation ditch and into the field. Occasionally, a mountain lion passes by, or a coati. More bullfrogs plop in the water as I cross back over the footbridge. More lizards skitter in fallen leaves. I'm aware of presence.

After a few weeks, I no longer use the charts provided by Nature's Notebook. I know what to look for and jot that information into a real notebook, one with good paper and a sturdy trim feel. I write down leaf size and reminders to myself: "Look up soap trees." My middle-aged brain has trouble absorbing new information, like the identification of an unfamiliar plant, and I notice how the trees seem to blur into one general category. Botanical dyslexia is something I've always suspected but I never had an official diagnosis. So I draw leaves as a way to remember them. Hackberry leaves are described as simple (a single leaf growing from the stem), alternate (each leaf has its own node rather than sharing a node with an opposite leaf), ovate to lanceolate (egg-shaped to lance-like), mostly serrate (saw-toothed along the margin), and sandpapery. The Arizona walnut is compound (more than one leaf growing on a smaller stem from the main stem), also alternate, lanceolate, and serrate. I sketch the triangular shape

Javelina (photo by Elroy Limmer)

of lambsquarters, prolific in my study area, and spend a few minutes serrating the leaf margins of the exotic Siberian elm, comparing that with the native American elm. I make some observations about wolfberry, which will later have green tubular blossoms and bright red fruit. I draw to size the tracks of javelina and deer and document the growth of sacred datura, every part of this plant poisonous and capable of causing hallucinations, blindness, and death. Yes, I'm keeping a naturalist's notebook, which fits in my pocket like a smartphone.

In Nature's Notebook, you can monitor multiple sites, and I add one more—a ten-minute walk to the Gila River. My new plants include American plum, Goodings willow, Fremont cottonwood, catclaw acacia, and rabbitbrush. My animals are the punctured tiger beetle, red-winged blackbird, killdeer, and great blue heron. I also watch for the pair of nesting black hawks, endangered in New Mexico but common elsewhere. Not all these species are on the preferred list. I just want to know them better.

I keep my eyes open for any small circular holes made by the larvae of the Western red-belled tiger beetle. We have had an odd spring weatherwise, dry and cold, with freezes well into May, each freeze followed by sunny days, hot, then cold, then hot. If I knew where these beetles had laid their eggs, where those eggs had hatched into larvae, I'd go feed them mini-mealworms myself, all thousands and thousands of those burrows with hopeful first or second or third instars anchored inside, patiently waiting for food to pass by. What kind of food is passing by and can it possibly be enough? It's hard to believe any of these beetles will survive in the wild without me.

* * *

The Upper Gila Watershed Alliance is a small grassroots environmental group to which I've belonged for fifteen years. This spring we hold a forum on mountain lions, with a panel made up of the state game and fish department's big-carnivore expert, our local game and fish officer, and retired biologist Harley Shaw, author of *A Soul Among Lions: Cougar as Peaceful Adversary*. I volunteer to moderate the panel, nervous that the audience might need managing—with opinions that range from the rancher's dislike of anything that could harm a cow to the anti-hunter appalled that mountain lions are game animals legally "harvested." When I meet the state's big-carnivore expert and the local game and fish officer in the parking lot of the Senior Community Center, however, I relax. One is a burly guy with a pierced ear and tattoos up and down his arms, the other taller than a redwood tree topped by a cowboy hat. Clearly neither will need my help. They have faced tougher crowds, and they focus steadfastly on the facts, with charts and figures and nice photography.

Adult lions weigh between 80 to 180 pounds and average seven to eight feet long, including tail. The cubs are born in the spring and stay with the mother about eighteen months before the sons disperse to find a new territory, preferably some hundred square miles. Daughters tolerate a smaller area, which may be close to where they were born. Although attacks on humans are rare, mountain lions can become habituated to human environments and conditioned to feed on pets and livestock. Here in the Gila Valley, domestic animals not securely penned are attracting these predators.

Mountains lions, in general, are not threatened by human activity. New Mexico has an estimated population of 3,000 to 4,000. As a game animal, some 244 were shot by hunters in the season of 2011–2012, well within a sustainable mortality. Lions also regularly kill each other. In a ten-year study of an unhunted lion population in central New Mexico, half of lion mortality came from other lions, usually males killing cubs or males killing females defending their cubs or males killing other grown males.

The fifty people who come to the forum want to know when they can legally kill a mountain lion, what the signs of an attack are, what to do when attacked, how many lions come down the Gila River corridor each year, are there really black mountain lions, why the state uses the word "harvest" instead of "kill," how long lions can live in the wild, and how to prevent lions from becoming habituated to their home. The questions are reasonable, and the audience listens politely to the answers.

Later in the spring, three more lions are killed in the Gila Valley by the game and fish department, a mother and daughter for eating chickens left in a moveable and unroofed pen in a field, and a male lion for eating a sheep. A few people in the valley had earlier seen the mother, shepherding her two adolescents, their tawny coats blending into rock. A friend of mine described how the mother's head had turned, slowly, slowly, watching the watcher, her expression unfathomable.

✳ ✳ ✳

In my daughter's classroom, the children are restless, more talkative, less able to stay in a routine. Summer is around the corner, although that corner is still a few weeks away. The tests that evaluated their academic progress for the year were finished over a month ago, and with the tests gone, in a system where learning is equated with testing, the point of learning seems to be gone too. The children sense this, although the teachers pretend it ain't so.

I bring in a citizen science project from the Cornell Lab of Ornithology. Celebrate Urban Birds asks children and adults to choose an urban area—the middle of a small border town like Deming, New Mexico, qualifies—and watch bird activity for ten minutes. Observations of certain birds—house finch, house sparrow, barn swallow, cedar waxwing, brown-headed cowbird, European starling, oriole, American robin, killdeer, mourning dove, rock pigeon, peregrine falcon, American crow, mallard, or black-crowned night heron—are recorded on a checklist, either online or with a paper form. Cornell Lab has sent me color posters with descriptions of these birds, a larger poster that shows their silhouettes, packets of flower seeds to plant, and stickers that say "Zero Means a Lot!" This theme is repeated a number of times in their instructions. "Send us your observations EVEN IF YOU SEE NO BIRDS in your bird-watching area. *Zero means a lot!"*

We head out with a gaggle of children to the playground where we face a row of planted conifers and deciduous trees, the school fence just past the trees, then a street and row of residential houses. The third-graders divide into groups of seven, each with a supervising adult, each with their own area to watch. This doesn't last long, with small boys running back and forth under the trees, and then entire groups dissolving and mixing. An American robin poses on a branch and puffs out its red breast. Rock pigeons swoop through the bare yard behind us. Mourning doves call from nearby telephone poles. A house sparrow lies dead on the other side of the fence, and this attracts far more attention than the live house sparrow in the tree. Some of us spot what we think is an American crow across the street. Some children also see peregrine falcons and Baltimore orioles colored red and yellow. We say no, probably not, and if they are stubborn, we say—okay, check the box that says "unsure." Some children remember birds they have seen before, the mallard at the zoo with a broken leg and the mean parrot kept by their grandmother. They watch for ten minutes and then come inside and learn how to fill out a form, a description of the site and their observations, carefully copying what my daughter writes on the chalkboard. I have an epiphany: this last activity (filling out forms) will be as useful to them as anything else they've done today.

Pleased, my daughter and I plan to do multiple Celebrate Urban Birds counts all next year. We'll do art projects about birds. We'll have graphs and word problems. We'll hand out more information about the common species in Deming, pigeons and grackles and Western kingbirds. Eventually these children will say, "I learned how to bird-watch in the third grade." Or, "I became passionate about birds in the third grade." Or, "My teacher's mother came into my third-grade class and revealed the world to be a web of connection and light, and I have never been the same since."

I may be getting ahead of myself.

In their promotion of citizen science, the Cornell Lab of Ornithology says somewhat ponderously, "Evaluations have shown that in addition to learning science content, many participants discover that

Mourning dove (photo by Elroy Limmer)

scientific investigation involves careful observation, adherence to data-gathering protocols, cautious conclusions, and a requirement for further investigation."

This implies any number of things. First, that learning about how science works is an important outcome. Second, that not all participants learn this.

I'm more than a little concerned I might fall in this category. In just the first month of doing Nature's Notebook, I've been surprised at how many mistakes I make—and how tempted I am to fudge the data. Behind my do-nothing yucca, for example, another yucca starts growing its stalk, a few more inches every day, swelling into a bud that will soon burst forth with delicious-looking, creamy-white, thick-walled blooms. I lust for that bud and then realize that the flowering yucca is a branching stem from my own yucca, the base of the mother plant having fallen on the ground and been covered with debris. Now I must go back and change the information I put into the online form. I also change the identification of some white-winged doves I thought were mourning doves. Later I realize the acacia I have been observing is not catclaw acacia, the species listed by Nature's Notebook, but Eastern white thorn acacia. I delete that plant, which involves sending an e-mail to Nature's Notebook explaining why. Later still, I misidentify the tight clusters of new leaves on the rabbitbrush as flower buds. I *blush*. I've been living next to rabbitbrush forever. How could I mistake leaves for flowers?

Moreover, the point of Nature's Notebook is to help document climate change, but because our land in New Mexico is in a cold air drainage in a river bottom, we have cooler temperatures than areas slightly higher, just up or across the road. My honey mesquite has not leafed out when so many in the Gila Valley have. My desert willow is bare while my neighbor's desert willows have already erupted in pink and lavender orchid-like flowers. My plants are weeks behind other Gila Valley plants and–oh no! I think, this will throw off the worldwide data! Now I can't be among the harbingers of doom! Would it really be so bad if I changed my times or my plants? Who would notice? Who would care?

And *another* thing: it's boring to input data. Scientists often complain about this and now I feel their pain. The form is easy enough. Mostly I check boxes. Mostly the information is repetitive. Not that much changes in three or four days. Rather soon I'm tired of checking boxes that never change and delete all the animals I am not likely to see. Now I am only noting the presence or absence of hummingbirds, cliff swallows, killdeer, herons, red-winged blackbirds, and punctured tiger beetles. (It's still too early for these adult beetles, a summer-active species.)

This May, Nature's Notebook reaches its millionth observation when Lucille Tower from Portland, Oregon, records her vine maple flowering. "Being a scientist is a frame of mind," Lucille says, "not a credential."

I don't have either, but apparently I'm not letting that stop me.

✳ ✳ ✳

In May, too, lightning starts a fire on Mogollon Baldy in the Gila Wilderness, with another fire in Whitewater Canyon, several miles to the west of Baldy. The area is rugged and steep, high winds cause the fire to spread, and by May 30, the merged Whitewater-Baldy Fire–the largest recorded fire in New Mexico's history–is less than ten miles from my front porch. Clouds of beautifully colored smoke bloom from the horizon, and at night we see flames–the crowns of trees exploding. During the day, smoke settles in the valley like a dismal fog, the sun a blood-red coin in the sky. The smoke and sky weigh us down, a depressive reminder. Climate change? The future is now.

Summer is almost here, and Barry e-mails with new ideas for my second field season.

From: C. (Barry) Knisley
Sent: Tue, 28 May 2012 10:29:33 AM
Subject: RE: western red-bellied
Sharman: In coming up with a project which I guess will
be at your nearby Gila R. site, are there any other species
there besides C. sedec and C. ocellata? As I mentioned, field
observation in a systematic way to determine daily activity is
useful and perhaps trying to determine if these two species
(any others there) exhibit any microhabitat segregation along
the water edge: establishing grids or transects along the width
of the shoreline and counting numbers of each species within
each section at intervals throughout the day is a way to do this.
Preliminary observations might indicate if there are differences,
interactions between the two. Repeating short time spans of
observing what the two species are doing could expand this
study.

Of course, the key thing would be resolving the field oviposition
sites for C. sedec. Observations probably later in the day
or at various intervals should be able to confirm if they are
ovipositing in the area where adults are active and if not that
would be good to know. Knowing what they do at night would
also be valuable. Possibly ocellata is ovipositing in adult habitats
and sedec is not.

Another aspect of resolving the puzzle of sedec oviposition
would be to examine the ovaries of adult females at periodic
intervals through the season. If they are not ovipositing during
their early season activity but rather doing it after they disperse
at the end of the season, we would not expect to find mature eggs

in the female ovaries until late in the season. Dissecting females
is not too difficult with a binocular scope, though not the most
pleasant thing. Barry

Looking at how the Western red-bellied and ocellated tiger beetle interact and behave differently at the river's edge does sound interesting. Grids
and transects. More play time in the sand. But since the ocellated didn't
appear in any numbers until last August, I decide instead to focus on that
"key thing," resolving where female Western red-bellied tiger beetles lay
their eggs. Part of that answer might lie in when they lay their eggs. If those
eggs don't mature in the female ovaries until late in the season, late July
or August, then the tiger beetles likely wait until after their dispersal to
upland areas to oviposit. To prove that hypothesis, all I need to do is collect
beetles through the summer and dissect their ovaries.

How far I have come–to think that so casually.

I am also e-mailing with a woman who is the first person to document a
tiger beetle in Tasmania, where we once thought that none existed, just as
we think they don't exist in Hawaii, the Maldives, the Antarctica, and the
Arctic north of 65 degrees. A high school teacher in Hobart, Tasmania,
Kristi Ellingsen first started using the photo-sharing site Flickr to post
macro shots of odd insects she found on vacations at the beach or just
around the neighborhood.

"I'd post a mediocre photo of some obscure fly," she writes me, "and
by the time I had woken up the next day there would be a conversation
between people I had never met narrowing down the wing venation until
they could tell me what species I had observed. I was hooked. I could be a
Charles Darwin: explore my backyard, and find exciting things. My kids
get dragged along and have now each discovered an undescribed species.
My middle son found a spider, my eldest a fly, and my daughter a bug."

The Ellingsen family seem to be on a roll, and they are not alone. The
number of described species on the planet is about 1.9 million, with some
19,000 new species described every year. Most of those are from the
tropics, and most are insects and plants. At the same time, even heavily
studied parts of the world like Europe are uncovering new species–770 on
average each year–and over 60 percent of these are being found by citizen
scientists.

On one beach holiday, Kristi photographed a large metallic beetle with intimidating jaws. Once before on Flickr she had mentioned finding a tiger beetle, only to have her new friends scoff: tiger beetles don't exist in Tasmania. This time she did a broader web search, came across David Pearson, e-mailed him three of her best pictures, and went to bed. The next morning she had a reply and then a confirmation that this was a species of tiger beetle, *Myriochila semicincta*. David suggested she write a paper for the journal *Cicindela* and prompted her to ask new questions, such as whether this beetle was a vagrant or part of a colony, and if its behavior would be similar to the same species living elsewhere, or could it, perhaps, be a sub-species?

David was doing what David does: prodding new research.

Black-tailed rattlesnake (photo by Elroy Limmer)

June 2012

June 3

Western red-bellied tiger beetles dot the riverbank, where I saw them last summer in mid-July. They are not yet in abundance, only three or four beetles per square two by two feet, and their behavior seems subdued. Smaller insects run across the sand, through the sparse grass and sedge, under a tiger beetle's very nose or glabrous labrum, and still the beetle does not seem to respond. This sluggish behavior might be a projection since I am feeling sluggish myself. Tiger beetles so soon? That means I should get my terrariums and collecting nets from the garage; dig up, sterilize, and screen dirt for those terrariums; order some mini-mealworms pronto; buy alcohol and small glass vials for the specimens I plan to kill, preserve, and dissect; and go out right now and catch tiger beetles. I need to spring into action. Immediately I go home and e-mail Barry Knisley.

June 4

> From: C. (Barry) Knisley
> Sent: Mon, 4 June 2012 11:19:34 AM
> Subject: RE: early
> Sharman: Thanks for the message. Seems quite surprising to
> me they have emerged already; not sure we actually know their
> seasonality so it's good you documented that. It might be they
> are early this year like lots of things, but that is really much
> earlier than I would have expected. I will get back with you in a
> day or two about methods for preservation and dissection. Barry

At Bill Evans Lake, I choose the loam from the same upland area at the eastern base of grasses that I used last year in the terrariums where the females laid eggs. No tiger beetles skirt the lake edge. I go back to the riverbank and collect ten beetles in the hope of rearing more larvae to describe. The Tai Chi stalk, slap of net, and duck-waddle-hop is something

115

lodged now in my physical memory. Stalk, slap, hop. Like riding a bicycle. I don't have time to collect ten more beetles to kill and preserve, but I'm not worried. Tiger beetle females need a period of "gonad maturation" after they emerge from the pupae. This period can last up to three weeks, and so these females probably do not have developed ovaries to dissect. I'm also dithering since I'm not sure what strength alcohol to buy or if I have to use instead the noxious formalin.

Late in the afternoon, Barry sends another e-mail:

> From: C. (Barry) Knisley
> Sent: Mon, 4 June 2012 5:19:42 PM
> Subject: RE: RE: early
> Sharman: I was trying to find my notes on preserving beetles for ovary examination, which I did years ago. Mostly, I have dissected them fresh. In the meantime, freezing may be OK. I will try some dissections of some I preserved in alcohol (possibly the 91 percent would work best) to see what ovaries look like. I can describe the dissection by e-mail; it is fairly easy. You will need a dissecting scope at about 20x, fine forceps, and fine scissors. Barry

Our working hypothesis is that these insects will have mature eggs, ready to be laid, late in the season–in August–just before they disperse from the water's edge. This is based on observations at lower elevations in Arizona, where Western red-bellied tiger beetle adults also appear early in the summer and congregate around drying water sources. The beetles disperse after the rains have started and appear at lower densities in the uplands. Barry and David could never find their larval burrows around those drying ponds and ditches, and in their joint paper about larval tiger beetles in Sulfur Springs Valley, Arizona, a discerning reader can sense their irritation: "Although *C. sedicimpunctata* is perhaps the most common and widespread species in the Valley, no larvae were found in the field and only one was obtained by rearing."

June 9

The Whitewater-Baldy Fire in the Gila National Forest has spread to over 260,000 acres. More than one thousand firefighters are on the ground, including thirteen hotshot and eight other crews, with sixty fire engines, twenty-eight water tenders, and seven bulldozers. In steep canyons, the heat and flames are so intense that trees and soil virtually incinerate. Thousands of acres of high-altitude spruce and fir will never recover. For larger areas, the damage has been less, mainly because these parts of the national forest have burned before, clearing out the underbrush and reducing the fuel load. Low-intensity fires every two to fifteen years are good for a ponderosa pine forest, where the trees can live to be five hundred years old, protecting themselves with thick bark and deep roots from surface fires. Periodic fire also enhances the pines' survival by burning out competing species and maintaining well-spaced trees that discourage intense crown fires. Other plants in these forests have adapted to fire, with seed germination stimulated by high soil temperatures, and with oaks, junipers, aspen, and willow sprouting rapidly after a burn.

For now, the state's largest fire in recorded history is on its way to being contained, and the new concern is flooding. When the monsoon rains come this summer, they will pummel the denuded ground, carrying soil and ash, causing mud flow and landslides, accumulating force and speed and sweeping through the drainages of Whitewater Creek and Willow Creek, through the small towns of Glenwood and Mogollon, lifting up homes and tossing sideways the single bar or grocery store. No one yet has died in this fire, although uncounted numbers of animals surely did. The real threat to human life and property is ahead.

June 10

I am on the river killing tiger beetles instead of selecting them for a privileged life in my terrarium. No longer a benighted game-show host—*you're a winner!*—as with the mini-mealworms, now I am Kali. Now I am Death. Rather than transfer the beetle from a pinched fold in the net to a small collecting box, I bring an uncapped glass vial filled with 91 percent alcohol up into the net, closer and closer, releasing the pinched area and trapping the beetle swooning with fumes against the netting, so that the insect falls into the liquid as if by choice, struggles briefly, and sinks inert to the glass bottom. I collect three or four tiger beetles per vial, their bodies tumbling against each other, wings open, genitalia extruded. Killing is easier than collecting live specimens, and the work goes faster, which is good because it is much less fun.

In *Tiger Beetles: The Evolution, Ecology, and Diversity of the Cicindelids*, David Pearson writes:

> Try catching tiger beetles and you quickly learn that they are past masters of eluding enemies. Not only can they run quickly, but they seem to sense your moves and fly away as you are in mid-lunge. When they land, they do so facing you with those big eyes, almost as if they were daring you to try again. Although their abilities to run and fly quickly from danger are notable, tiger beetles face much more sophisticated enemies than tiger beetle collectors.

It makes me feel better. I am not the only doom stalking the bank. More sophisticated enemies include large, long-bodied, long-winged robber flies that capture tiger beetles in midair and kill the dangling prey with an injection of poisonous enzymes. Birds like flycatchers and jays eat tiger beetles, as do some spiders, scorpions, and predatory bugs. Which predator eats which species of beetle depends on the beetle's defenses.

The Western red-bellied tiger beetle flashes an orange abdomen in flight, a warning signal that this insect gives off the toxic chemicals benzaldehyde and cyanide. (Sometimes those excretions leave brown stains on my collecting net, although they rarely reach my nose and never my tongue.) To test whether that color actually deters robber flies, David Pearson set up an experiment in which he made papier-mâché models of tiger beetles tied on strings attached to a wooden pole. Dangling these in front of a robber fly, he predicted that those models with rear ends painted orange and dabbed with a drop of benzaldehyde would be less attractive. He was right. A larger-sized model resulted in even fewer attacks.

Then David wondered why all tiger beetles don't have orange abdomens (less than 10 percent do) and the ability to secrete a chemical like benzaldehyde (less than 40 percent can). The answer is that, unlike robber flies, birds and lizards are *not* deterred by a flashing orange abdomen or the defense chemicals. Nor is size a reliable defense. While some lizards and spiders require small prey, shrikes and flycatchers relish something

Robber fly (photo by Elroy Limmer)

bigger. What helps you escape one predator leads you into the claws and teeth of another.

The enemies of tiger beetle larvae—those pale kinked grubs anchored in their tunnels—can also be birds, such as woodpeckers hunting a meal, but more commonly are parasitoid wasps and flies. These wasps seek out a larva in its burrow hole; sting, paralyze, and lay a single egg on the grub; and then plug the tunnel. The larval wasp hatches, eats the immobilized tiger beetle larva, and emerges in a few days. Similarly, the parasitoid bee fly hovers outside a burrow hole and flips its eggs into the entrance; some of these eggs fall and roll to the bottom of the tunnel; the bee fly larvae hatch, crawl, attach themselves to the tiger beetle larva, and start eating when the tiger beetle larva reaches the pupa stage.

Like Sigourney Weaver in the *Alien* films, the Western red-bellied tiger beetle fights back. Feeding as larvae in the spring and emerging as adults in early summer may reduce some mortality by parasitoids. The adults are armed with cyanide and a flashing orange warning sign. The beetle is too small to be of interest to some bird and lizard species, and congregating in large groups may have the herd advantage of multiple eyes on the alert for predators. Of course, both the tiger beetle larva and adult are fierce predators themselves, taking the role of murderous alien as well as the victim whose body is an incubator.

Horror movies grab our psyches with that primal truth, our relationship with the world, our relatives and family. Hello? How are you? I think I'll lay my eggs in you. I think I'll eat you. The mountain lion turns her attention to a deer or goat or, rarely, a human child, thigh muscles bunching. The blue jay and the robber fly, too, are dispassionate predators.

Yesterday, while on my Nature's Notebook walk to the river, Peter and I got the attention of a prairie rattlesnake, an excitable species, which coiled and rattled and rattled and rattled, defending its territory and philosophy of life: don't tread on me. These days, we rarely encounter animals who can kill us. With my close-focusing binoculars, I made a quick identification– those chocolate-brown squares down a yellow-brown back, the triangular head with dark bars from the golden eye to the corner of the mouth, the black elliptically vertical pupil, and that sound in the air like a plane landing too close. Afterward I stayed on just to watch the tongue flicker.

Nature's own notebook is a ledger of who is eating and being eaten, birthing, dying, with humans adding eccentricity to the mix. I am not

Western red-bellies in glass vial (photo by Sharman Apt Russell)

killing tiger beetles in self-protection or for calories or to reproduce. I want to know where the Western red-bellied tiger beetle lays her eggs. I am filling in the blank spaces on a map. I collect and watch the insect struggle and sink to the bottom of the glass vial. I am Kali but conflicted, mentally aghast, and then I mentally shrug and collect the next beetle. I suspect this is what intrigues me most—what so often compels us, the study of ourselves—that deft movement from aghast to shrug.

June 17

While I am in Los Angeles teaching, as I do every June at a low-residency MFA program, Peter collects and kills tiger beetles for me. He is Death. He is doom. Thank you, I say when we talk by phone.

June 24

Peter and I follow our biologist neighbor, Mike, his biologist wife, Carol, and their nine-year-old son over lichen-colored rock and fallen trees tangled with grapevine in a narrow canyon in the Gila National Forest. The scale here is intimate, nothing grand or oppressive, the original feng shui.

We exclaim at patches of yellow columbine, long petal spurs and long stamens reaching into the air in slow-motion explosion. We exclaim at the baby birds in the nest of a willow, their beaks open for food, their pink bodies piteously ugly. Douglas fir grow in this rocky bottom, and more fir and ponderosa pine ascend up the canyon slope, the sky a river slowly darkening above our heads. This is one of those secret places holding water tucked into a waterless expanse of forested hills and bare-bone cliffs, a place where you might feel that you are the first person to ever climb this lichen-covered rock or see this pool of summer rain.

For the last six weeks, Mike has been camping in the national forest studying the Mexican spotted owl. From dusk to dawn each night, he hoots, waits for a return cry, and follows the sound. In this way, he has confirmed seven breeding pairs spaced apart in an area of about ten square miles. Mexican spotted owls are one of the largest owls in North America, with a wingspan of forty-five inches, famous for requiring old-growth forests of white pine, Douglas fir, or ponderosa pine, forests with high, closed canopies and tree cavities or holes in cliff ledges for the owl's nest. In the United States, an estimated twenty-one hundred Mexican spotted owls remain (with far fewer in Mexico), almost all on public land in New Mexico and Arizona. Mike is getting paid by the Forest Service to know where some of those owls live, find their nestlings, and observe their success at raising a family.

Just ahead of us, he carries a box of white laboratory mice. The protocol is to offer four mice, one at a time, to a breeding pair of Mexican spotted owls. If an owl swoops down, grabs the mouse, and eats it on the spot, she or he probably does not have hungry nestlings nearby. If the owl swoops down, grabs the mouse, and takes it away, she or he may have young that need feeding, and if you follow that owl and suddenly flying mouse to wherever they are going, you may be lucky enough to find where that is.

This all sounds pretty improbable to me. But Mike is a gifted naturalist and knows that this canyon is frequented by a particular breeding pair: iconic chestnut-colored birds, white and brown breast feathers ruffling in a mottled pattern, dark eyes bottomless, short curved beak a comic afterthought, expression singular. Owls look at us, and what do they see? Are they always this judgmental?

"That's a good sign," Mike turns and says, pointing to a splash of white feces on a rock.

I'm excited that a Mexican spotted owl has been here and might still be close by. This species has a resonance, long associated in the American West with litigation and the protection of habitat from logging and overgrazing. The presence of these birds indicates a healthy forest. Their absence is usually related to human impact, as well as drought, wildfires, and predation by other owls. In the Southwest, the Mexican spotted owl–listed as threatened under the Endangered Species Act in 1993–quickly became a scapegoat for an unsustainable logging industry already in decline (much like the Northern spotted owl in the Northwest). The controversy got heated, inspiring bumper stickers that I remember: *I like Spotted Owls–Fried.* Or *Save a Logger–Kill a Spotted Owl.* Under new federal regulations, logging practices changed, and eventually the anti-owl stickers were replaced by other rural concerns. Today the danger to these owls is all about climate change, a drying forest, and degraded habitat.

"Neat," I say, and then, "Oh, Mike!" because right behind him, perhaps twenty feet away, those big round eyes in that disc-shaped face are looking down at us from a pine branch. We all stop and gape and stand

Mexican spotted owls (photo by Mike Fugagli)

back, waiting for the bird to fly away. Soon, like incompetent paparazzi, we're fumbling for our cameras.

"These birds are so fearless," Mike stage-whispers. "I've never worked with a species that cares so little about the human presence." In the next two hours, we will watch the female swoop down and take two of the live white mice that Mike's son places on a carefully selected spot, usually the top of a jutting tree branch from which the mouse cannot easily climb down. The mouse seems puzzled but not alarmed to be out of its box, whiskers moving, sniffing at this brave new world. (I have some feelings for this animal, too, which I suppress.) The female owl swallows the first offering whole, the gulp and throat movement both casual and awkward. The male joins us within half an hour and also swoops down for two of the mice. It is as though this couple has welcomed us into their living room, hospitable hosts. Hello, hello. They swivel their heads. They eat our canapés. They hoot. They stare.

Then the female flies with her second mouse down the canyon to a creviced cliff face, and suddenly we are scrambling after her, running back the way we have come. Full stop! Panting! We hope that this rocky crevice, about thirty feet up, shelters her nest. Alternately, this may simply be a cache, a storage area for white laboratory mice. Since we can't see or hear any young birds, Mike suspects the latter.

The river of sky has darkened to cobalt-blue, and it's time to go home. Except for the nine-year-old, who has grown up with these experiences, we feel like we've won the lottery—maybe not a million bucks, but a few thousand!—giddy with this unexpected gift, privileged to spend these hours with Mexican spotted owls, environmental celebrities.

On the hike back up from the canyon to our car, we carry flashlights, and Peter spots the gleam of a Sonoran mountain king snake, banded black, red, black, yellow, black, red, black, yellow, black. For the nine-year-old's benefit, the adults chant the childhood rhyme, "Red on black, friend to Jack; red on yellow, kill a fellow," distinguishing the harmless king snake from the poisonous coral snake, banded black, yellow, red, yellow, black, yellow, red. Even as I join the others, I'm remembering that rhyme differently. Wasn't it "Red on yellow, friendly fellow?" This is the grave problem of mnemonics.

June 28

For a long afternoon, I watch tiger beetles. I have always wanted to be John Burroughs, imagining Zen-like moments parsing out a leaf, hours and days that pass like a dream, sun-kissed, plant-besotted, imagining, like so many others before me, a kind of rapture in nature and loss of ego. I see the beetles bask, pressing their lower bodies against warm sand. I see them shuttle back and forth between shade and sun, dry sand and wet sand. Late in the day, as the sun shines too brightly, I see them stilt, extending their long legs to get above the warmer boundary of soil and air. I see them face the sun, orienting their bodies so only their head is fully exposed to solar radiation. I watch when they fly: long flights create heat from muscle contractions, but short flights reduce heat through convection. In one study, David Pearson and Barry Knisley estimated that tiger beetle adults spend 56 percent of their day fine-tuning how hot they are.

I don't see a lot of mating, not the frenzied sex of the end of last summer. I don't see any females laying eggs. I do see ants and spiders and other unidentifiable insects, but I don't observe any successful hunts, although the beetles regularly dash, dart, dash, and then stop. Run, pause, look.

With their big eyes, tiger beetles can spot the movement of small objects, but when the chase has begun and the object is running, too, the beetle has to periodically relocate its prey. Run, pause, look.

As well as good sight, tiger beetles have ears, or a hearing tympanum, two domed structures on top the first abdominal segment. (Ears on their stomach. The Picasso world of insects.) But hearing is probably used more to escape predators than to find food; when a tiger beetle hears the ultrasound of a bat, for example, it contracts its abdomen and tumbles to the ground.

Tiger beetles can also smell and feel, and one species—the giant *Manticora redux*—seems to use smell more than sight while hunting. We don't know much more than that.

Nothing comes to prey on these beetles. Perhaps my presence discourages birds, and robber flies are not out in force. That will come later in July and August, when those powerfully built insects helicopter the river like a military invasion. For now the most formidable predator here is right in front of me, mandible-scything, digestive-juice-dripping. Run, pause, look.

The sun shines. The plants besot. Willow, sycamore, cottonwood, mullein, grapevine. Everything in its place and exploding with energy. Rabbitbrush, sweet clover, four-o'clock, spike dropseed. Run, pause, look. I'm open to insight. Ready for rapture.

But for every plant I know, there are so many I do not. For every familiar bird, there are so many strangers. Nature's Notebook has schooled me in my ignorance, and suddenly I'm brooding on that, surrounded by a world I can't name, fail to see. The blind voyeur. The autistic. A world, moreover, that is rapidly getting drier, warmer, a world that may look quite different in thirty years. Run, pause, look. What would John Burroughs do? The sun is hot. My lunch is eaten. The Jane Goodall Institute has never written back either. The best I have gotten is this brief e-mail: "Thank you for your interview request. We will reply as soon as possible." Then dead silence for the rest of my life.

Heavens, we are at the mercy of our moods.

June has been a discombobulating month, dreary with smoke and fire and the fear it will never rain again. I remember this fear from other Junes, although this one feels like a longer negotiation. The Gila River is

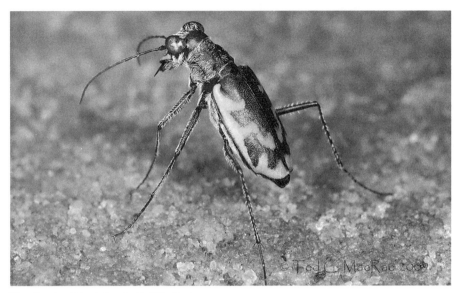

Tiger beetle stilting (photo by Ted MacRae)

historically low—the water in front of me more of a stream, something like a creek, a beautiful wet ribbon. What we have come to expect historically is history, and as the American West heats up, ecosystems will shift, forests to shrubland, grassland to desert. Less rain and snow, more snow melt in the winter, less water in the ground. More evaporation. Less moisture in trees. Bigger wildfires. More flash floods. More dust storms. More time fine-tuning how hot we are.

I know what this means. You can be angry, and you can be sad, but in the end you just have to love harder. Love the river. Love the desert.

June 29

As I walk home with my collecting boxes, a coati crosses the dirt trail about fifty feet ahead, angling through a field of buckwheat, tumbleweed, and grama grass. Typically a lone coati would be male since the bands are matriarchal, with the males leaving the group when they reach sexual maturity, returning in the spring for courtship and mating. But in early to mid-June, pregnant females also go off on their own to bear young, which they will care for alone through the summer. This could well be a female then, returning to her den, her average of four kits recently born. At first, a kit's head and nose are grotesquely large on top its tiny four-ounce body;

its eyes will remain closed for another week to ten days. For about six weeks, the mother leaves her kits in their nest as she forages, returning throughout the day to nurse. In August or September, she brings them out to rejoin the troop. For another few months, they'll continue to nurse from her, as well as from other females, growing up strong, coddled, indulged by their clan for as long as two or three years until they are fully grown adults.

This animal's reddish-brown coat looks thick and healthy, with a creamy white bib on the chest, something I see as she turns to stare at me. The lighter-tipped hairs on her back and shoulders give her a golden glow, and like many coatis, she does not seem alarmed by my presence. This lack of concern is typical even of coati mothers, and Chris Hass, a coati expert, speculates that the nonchalance could be "decoy behavior," with the animal deliberately lulling and leading potential predators away from her den.

Chris Hass is one of a very few people who has studied coatis in North America. Another is sportswriter and author Bil Gilbert. In 1970, Gilbert spent a year watching two bands of coatis in southern Arizona, accompanied by his seventeen-year-old son and two other adolescent boys. Camping out at first, and then settling in a stone cabin in the Huachuca Mountains, they logged in 290 days, 8,500 observation hours, and 5,000 recorded sightings, some brief, some lasting hours, some for long afternoons, sun-kissed, plant-besotted. In hindsight, they made serious mistakes, feeding the coatis dog food and marshmallows, keeping the troop artificially around the Huachuca cabin, and later–some biologists suspect–losing or at least never writing up their field notes beyond Gilbert's *Chulo: A Year Among the Coatis,* published in 1973.

Even so, they saw extraordinary things. They had intimate glimpses into the life of a coati band, the ritualistic and hygienic grooming ("a sensuous, luxurious experience … at times it seems that a grooming pair has entered a trancelike state"), the complex relationships among females (the two leaders of one band, Queenie and Witch, were fast friends, while Witch and Calamity bickered and fought), and the way the males circled the matriarchal band for seven or eight months of the year like "orbiting satellites." They watched coatis cuddle, fight, mate, die, and birth. They personalized the coatis, much like Jane Goodall did her chimps in the forest of Gombe, using the anthropologist's intuition–admittedly through the filter of another species. The Witch was "a small, old, toothless, partially blind animal of superior intelligence and experience. Her sense of smell was remarkable, and her personality was subtle." Calamity was a large aggressive handsome animal–somewhat more "impetuous and unstable than the other females." The Supervisor was a male "of great boldness and dignity." The Bungler was an "intelligent, inquisitive, but self-indulgent animal ... not enough of a warrior to become a lover." At the end of their year, Bil and his boys felt themselves closer to being adoptive members of a chulo tribe than any other human being in the world.

The phrase *green with envy* probably comes from the early Greeks, who believed that jealousy was accompanied by an overproduction of bile and a yellow-green complexion. In the seventh century BC, the poet Sappho described the face of a jilted lover as green, and Shakespeare used the same description centuries later. Although envy is one of Christianity's seven sins and universally abhorred, scientists now think there is *benign envy* as well as *malicious envy*, with the former serving as an important, perhaps evolutionary motivator. In one experiment, subjects who were revved up to be envious (first writing down their four most envious moments) significantly boosted their performance on tests of memory and intelligence.

To spend a year with the coatis. To look through that window into the mind of another species. To be an adopted member. To learn their language, at the root of which are three sounds: squeals, grunts, and chirps. In Bil Gilbert's research, a certain loud sustained squeal–*I am in terrible trouble*–evoked an immediate response from any nearby adult. Other squeals between two cubs fighting–*I want it! Give it to me! That's mine!*–were usually ignored. Adults often gave soft, absent-minded grunts–*All is well*–while foraging. A more focused grunt between two coatis meeting–*Hello*–resembled the word *chulu* to human ears. A snarling grunt–*Challenge!*–was quite a different noise. A trilling bird-like chirp seemed to indicate affection and contentment. And then there was the startling and specific *ha-ha-ha*, what sounded to these citizen scientists like amusement and excitement tinged with contempt, what the Bungler often used to announce his presence, what males did before bluff battles determining their hierarchy, and what females and males did before copulation.

As Gilbert reports: "One afternoon in late April, when feelings of romance were strong and general, the ravine behind the cabin sounded like a hysteric ward in an asylum as two males and four females, in various combinations, called *ha-ha* back and forth in maniacal fashion. When the tribal elders began to *ha-ha,* it seemed to alarm the cubs mildly, make them restless, like children in the presence of drunk, disorderly, or impassioned adults."

Reading Gilbert, I find myself turning green with benign envy.

Unperturbed, the coati continues her measured progress through the grass. I have time to get out my binoculars and follow her path, perhaps deliberately in the opposite direction of her den, until she is out of sight. She is looking for roots, lizards, fruits, nuts, and tarantulas, feeding herself and then producing the milk that will feed her babies, those soft, gray, big-headed, big-nosed lumps nestling against her and squealing at the slight-est provocation, squealing and rooting, wanting more milk, more touch, more, more, more. She chirps in return, "Pumpkin-cakes. I'm here."

June 30

From *The Gospel of Nature* by John Burroughs: "To enjoy understand-ingly, that, I fancy, is the great thing to be desired. When I see the large ichneumon-fly, *Thalessa,* making a loop over her back with her long ovipositor and drilling a hole in the trunk of a tree, I do not fully appreci-ate the spectacle till I know she is feeling for the burrow of a tree-borer, *Tremex,* upon the larvae of which her own young feed. She must survey her territory like an oil-digger and calculate where she is likely to strike oil, which in her case is the burrow of her host *Tremex.* There is a vast series of facts in natural history like this that are of little interest until we understand them. They are like the outside of a book which may attract us, but which can mean little to us until we have opened and perused its pages. The nature-lover is not looking for mere facts, but for meanings, for some-thing he can translate into the terms of his own life. He wants facts, but significant facts–luminous facts that throw light upon the ways of animate and inanimate nature."

Dragonfly (photo by Elroy Limmer)

July 2012

July 3

Traditional lore, pre–climate change, is that the rains start by the Fourth of July. This year that seems to be true. Clouds gather in the early afternoon, warm air rising from the heated ground and then cooling. The cloud's flat base is the level where condensation begins, while the rest of the air continues upward in stacks of puffy white, up through temperatures that are dropping rapidly, warm air meeting cold air as the cloud develops higher and more vertically with peaks and towering cliffs. Inside the cloud is further rising and falling, condensation, coalescence, until water droplets become heavy enough to fall. In cloud language, the word nimbus means rain, and a cumulus cloud has just become a cumulus-nimbus, perhaps seven miles in height and several miles wide. High altitude winds shear its top, the anvil from which trails of ice crystals or cirrus clouds spin out in fibrous wisps called mare's tails. Electrical energy builds up as water and ice particles are repeatedly split and separated. Suddenly there is brightness, flashing, cracks, and rumbles—a late afternoon thunderstorm. The humans rush to turn off their computers. The dark gray clouds release their swollen bellies, water falls from the sky, and the humans dance. Or maybe they just go about their business, in and out of stores, sitting at a desk, watching children, fixing a car, but suddenly happier. The rains have come.

Enough rain has fallen throughout the national forest that some people in government agencies get nervous and order an evacuation of the town of Glenwood, New Mexico, as well as the nearby smaller towns of Alma, Pleasanton, and Mogollon. More heavy rain is expected this weekend over the burn scars in Whitewater Canyon, with flash flooding and debris flow warnings, over two hundred thousand sandbags distributed, and evacuees reminded to pack up the six Ps: people, pets, important papers, prescriptions, family pictures, and personal computers. Residents have fun with the six Ps. Was that prunes, peaches, persimmons, parsnips, parsley, and pigs feet? Or Plato, Plutarch, Poe, Potter (Beatrix), Proust, and Paolo Bacigalupi? Those who don't choose to evacuate sign a waiver releasing the government from liability, while those who do head over to the high school gym of another town to spend the night, where–I imagine–they exchange tongue-twisters of the Peter Piper Picked A Peck of Pickles variety.

Typically, it's not raining at all here in the Gila Valley, although the sky is gray. My husband, friends, and I are also talking about debris flow, but for a different reason. An unholy alliance of Grant County, the Freeport-MacMoran Mining Company, and the Army Corp of Engineers have used emergency federal funds to bulldoze a channel in the Gila River near the Highway 180 bridge and the smaller Highway 211 bridge. Ostensibly this construction work is meant to control the water and protect these bridges in case of a flood. But unlike the canyons of Whitewater and Willow Creek, the Gila River is not in danger of unusual or severe flooding. Moreover, this river is a healthy system *meant* to flood, change and shift its course on the wide valley floor. Suddenly bulldozers have turned stretches of riparian willow and cottonwood into a gravel pit. This happened almost overnight. There's no putting back the trees now.

July 5

The Fourth of July weekend continues without rain where I live–although we see it over there. Over there, someone is getting wet. I drive over to the Mogollon Box Campground on the Gila River where Western red-bellied tiger beetles are skittering on the bank and hunting in the grass and sedge. After I collect and kill ten of them, I pause before a bush of flowering white clover, the plant fairly winged with so many pollinators. The monarch: bright orange edged with black like the panes of a stained-glass window. The common buckeye: softly brown like deerskin, its dark eyespots ringed in yellow. The fluttering cabbage white: whose caterpillar was found to measure daylight using pigments in the blood, able to distinguish between fourteen hours and fourteen and a half. The fluttering checkered white. The fluttering clouded sulfur. The fluttering hairstreak. The fluttering Western pygmy blue. And all the pairs of mating netwings, orange soft-bodied black-banded beetles with black, segmented antenna scattered on the flowering bush like confetti or, at second look, like couples making out at a high school party.

And the sound of water flowing with no end. No stopping point, the river sound, water over rock, water and water and water without pause or interruption, all day, all night, all day, all night. Perhaps it's the desert in me that finds this so stupefying.

Under the clover bush, the sweet smell and fluttering wings, a gray earless lizard defecates and runs away. Two ants come forward like waiting sanitation workers, officious with a broom and pan, to pick up the brown pellet. This makes me laugh and sit down and watch them. Ants are like the endless channels on cable TV. Something strange is always going on. Yesterday, Peter and I saw a little mound on our gravel driveway that turned out to be a pile of ant heads. There were other ant parts, but mostly heads, the remains of a meal left by a toad or horned lizard.

At ground level, closer to the water, I see a flame skimmer, a common Southwestern dragonfly, perched on a stalk of grass, making the grass bend, which makes the dragonfly shift his balance. The flame skimmer is recognizably male with that carrot-orange body, and I want to avert my eyes, initially dismayed by the hunched shoulder and round head that seems featureless, a man with a ski mask. Then the bulbous eyes emerge, almost the entirety of this insect's face, with a small mandible and labrum beneath. The dragonfly stares me down, prehistoric, knowing, sly, or maybe that's just me—which is always the case. We look at the mirror of nature and see ourselves, all that meaning and beauty, horror and brutality.

The scholar and Jesuit priest Thomas Berry popularized the idea of human consciousness as "the universe reflecting on itself," the Big Bang, birth of stars and planets, evolution of life on Earth and specifically of *Homo sapiens,* all resulting in a woman making a face at an orange dragonfly, revving up some drama in her life—a man with a ski mask!—and thinking about dinner. Leftover chicken? Mostly she's feeling smug that she lives in such a cool place, by a river, water over rock, water and water and water without pause. She's feeling special, surrounded by butterflies and dragonflies. She's so smart to be here.

Perhaps the universe could have chosen more wisely, but let's not spoil the moment.

* * *

A loud, tinny announcement from God interrupts, "If you are within the sound of my voice, you must evacuate this campground. If you are within the sound of my voice, you must evacuate this campground. If you are within the sound of my voice ..."

After that initial confusion–God's voice would be much deeper–the writing teacher in me responds next. Are people *not* within sound of that voice exempt from evacuation? Is it redundant to mention the sound of a voice? I've never heard of a riverbank evacuation before, but I am genuinely grateful that the Grant County Sheriff's Department cares enough to save my life in the event of a flash flood, something I know is more than possible. I gather my things, dead beetles in glass vials, and walk over to the cruising patrol car. The deputy explains that a sensor installed on the burn scar of the Whitewater-Baldy Fire recently measured two inches of rain in an hour, and that's reason enough to ask people to leave. I've noticed myself how brown the Gila River is running, dark-chocolate-brown with ash and soil. "Up there," the deputy points north to the fire, "some creek beds are thick as tar."

Back at my house, I see a familiar skitter and dart on the gravel path. A Western red-bellied tiger beetle, a quarter mile from the river. Are they dispersing so soon? That would make sense as the rains begin in earnest and their habitat starts to flood. "If you are within the sound of my voice," I tell the beetle, "you must evacuate."

July 11

At Bill Evans Lake, where the water level is stable, tiger beetles are everywhere, and I collect some for a second and third terrarium. Then I drive on to the Gila River Bird Refuge, white thistle-poppies along the dirt road, fields and meadows waving green, and the cloud-filled sky so magnificent I want to stop and show it to someone. I want to take a picture or video to post on YouTube. I want to use globs of oil paint and put the sky in a gilt frame. I want to slow down, slow down the clouds massing and lifting and colliding, slow down myself so as to better appreciate what I am seeing. Why am I studying tiger beetles? I should have been a cloud scientist. Categorizing clouds: low clouds like stratus and stratocumulus; middle clouds

Cloudscape (photo by Elmer Limmer)

like altocumulus, altostratus, and nimbostratus; high clouds like cirrus, cirrocumulus, and cirrostratus. And pileus! the bouffant effect of super-cooled droplets above a cumulonimbus; mammatus, a sea of hanging udders from an altocumulus; virga, dissipating ice crystals like the filaments of a jellyfish. I want to live in the clouds and feel them pass through me.

And, yes, there is a citizen science program for this. In the S'Cool Project, citizen scientists report their observations of clouds to help confirm the accuracy of NASA satellites orbiting Earth. Observations can be timed to the very moment a weather instrument is passing over and measuring the same clouds. The scientists at S'Cool explain, "The cloud properties we are seeking are cloud type, cloud height, cloud cover, and cloud thickness." This sounds hard, but they provide worksheets. Cloud observations are increasingly important as we begin to connect clouds with global warming, and as we struggle still with the most basic questions: how will global warming affect clouds, and what kind of cloud cover increases or decreases global warming?

What I really think: I should be on my knees. The sky is a religious landscape, not a scientific one. I think about my father, a test pilot who flew and crashed the experimental rocket-powered aircraft X-2 in 1956, going three times the speed of sound—briefly the fastest man on Earth. Mel Apt died when I was two years old, and although I don't know much about him, I do know that he loved clouds. On home movies taken over the Grand Canyon, he does not pause long over his wife and two daughters before he is panning that new 1950s movie camera across the clouds massing and billowing in the Arizona sky, clouds he knew from many hours of flight in all kinds of airplanes, crop dusters and F150 Starfighters and B-50s, clouds where he felt very much at home.

A home in the clouds. I park the car and walk down to the Gila River, where there are thousands, *thousands* of Western red-bellied tiger beetles swarming its sandy banks. I'm not exaggerating–I can count ten to twenty beetles every square foot, and I walk half a mile downriver (2,642 feet) and still see that same density, tiger beetles darting, skittering, flying away as I approach, landing and looking at me inquisitorially with their bulging eyes and scything mouthparts. So much life, the teeming world of insects. Some of the beetles are mating, but not as many as I might have expected. No one is ovipositing. (And they wouldn't, they shouldn't, on this sandy soil so likely to flood soon.) Mostly the tiger beetles just teem.

I walk and watch with my close-focusing binoculars, trying to spot any other species, like the ocellated tiger beetle, remembering Barry Knisley's suggestion that looking at how the two species share a habitat would be useful, whether they "exhibit any microhabitat segregation along the edge," and determining any "differences, interactions between the two." But it's only Western red-bellied tiger beetles, one after another like a conference of funny-hatted Shriners, this river booked for members only. You have to have the seven creamy dots.

I watch and walk and sometimes slog, slipping into the river, which is still running dark from the ash of burn scars thirty miles away. The clouds mass and tangle overhead. Perhaps we'll get an afternoon storm. The sun shines periodically through the extravagant cumulus and cumulonimbus and altocumulus, and I am open to adventure, ambling, wide-eyed. A striped whipsnake flows across the trail. A black hawk keens. A summer tanager is red, red, red. This is a childhood I didn't have, the wild and grubby boy or girl (but usually boy) roaming his private paradise, collecting treasure and bringing it home: hardened pupae, arrowheads, crumbling seed pods. Those free-spirited children filled their pockets, and I'm filling mine, clothes torn and leaking leaves. Feathers in my hair. Maybe I'll see a lion. Maybe an ocellated.

July 14

I'm avoiding dissecting tiger beetles. Cutting into the ovaries of small insects ... it's so hard to fit everything into the day. Helpfully, Barry sends instructions.

Dissection:

a. Remove female from alcohol and pin into the bottom of small petri disc or other small container with a pinnable substrate in bottom of disc. A one-inch layer of melted paraffin works well. Place beetle venter down and put one pin through the thorax and a second pin through the posterior portion of the ovipositor. See Fig. 1.

b. With fine forceps remove the two elytra and then the hind wings (remove these by grasping at the base of the wing). See Fig. 2.

c. Remove the dorsum of the thorax by inserting a small scissors along one side of the membranous lateral portion (between the dorsum and venter) and cutting anteriorly through the chitinous portion at the junction of the abdomen and thorax where wing is attached. Then cut laterally across to the other side around the junction of abdomen and thorax. Repeat this on the other side of the thorax. At this point the dorsum should be free except for the posterior end, and can be carefully pulled posteriorly to expose the contents of the body cavity (Fig. 3). Now ideally flood the specimen with 100 percent alcohol (or more preferably FAA, if available). The FAA hardens and fixes the ovaries so they can be preserved or mounted on slides and carefully separated to count the number of mature eggs. Most of what you see are two large white ovaries with the ovipositor and female genitalia between them. In Fig. 3 you can see the left ovary while the right one is partially obscured by the dislodged ovipositor. Examining the ovaries, if mature, will reveal a series of large oval eggs. If the ovaries are undeveloped without eggs, the two ovaries will be very small strands or nearly invisible.

The figures are instructive. As Barry wrote: "not too difficult...although not the most pleasant thing."

And then the actual experience is more horrible than I had even imagined. For so long now I've watched television shows like *Law and Order* and *CSI*, and their thousands of spin-offs and endless reruns until being a coroner or forensic pathologist—cutting up bodies, slicing and dicing, cracking rib cages, lifting up hearts—had begun to look like the most natural of jobs. I didn't understand that a body—any body—is whole, a whole creation, and that there is something unholy in taking it apart. Ripping off wings. Removing the dorsum. The smallest scissors I have are like using a chainsaw to cut someone's hair. Oops, did I get your head? Heads, in fact, are rolling across the petri dish and when the tiger beetle's back is lifted away, the insides come, too, and bits of wing and leg and antenna, and this is butchery, inelegant, demonic.

Moreover, I have the nightmarish feeling the beetle is still alive. Or will suddenly spark to life. Zombie beetles.

After a time, all this becomes normal, normal being relative, and unsuccessful. Cutting anteriorly, cutting laterally, pulling the dorsum posteriorly—each time, I destroy what is in the body cavity rather than reveal it.

I'm also having trouble finding females. Male tiger beetles have dense pads of setae or hair on their front legs, which they use to grab and hold onto the females while mating. For most of the beetles collected in June, I can see these dense pads clearly under the microscope. In a few cases, I am less sure. Male and female genitalia are different, of course, but that's a distinction I am not able to make, although I would recognize a developed ovary if I could manage not to obliterate it first. I start skipping around through the vials, looking for females from later in the season, and when I find what I think is one, I tear her apart, putting on my Dr. Frankenstein face, eyebrows drawn together. I want to do the same with the instructions. As easy as giving yourself a colonoscopy.

I decide to defer dissecting until Barry Knisley's visit.

July 21

Barry is coming with six former students, one of whom is part of a couple, and so I worry about providing six separate beds, six sheets, six pillows, six coverlets. I expect six former PhD students, a gaggle of entomologists, but in a last-minute e-mail I am reminded that these are men and women from Barry's first years of teaching undergraduates at Franklin College in Indiana, where the enthusiastic young biology teacher took his students on field trips, his interest in tiger beetles bright as a new dissecting tool. Barry went on to become a tiger beetle expert and his undergraduates went on to become public school teachers and computer technicians and administrators and one factory worker. Still, they kept in touch and, forty years later, are joining together for another road trip.

The afternoon before Barry's visit is a day of rain. The Gila River rises, not a big flood but big enough to cover all the areas once teeming with Western red-bellied tiger beetles. Although I had wanted to show off this abundance, I'm not really disappointed. Here in New Mexico, rain can disrupt almost anything–a party, a wedding, a one-time visit from a mentoring entomologist–and still be a good thing. Rain is almost always good.

July 22

Barry Knisley is trim, nut-brown, and laid-back, with an easy handshake and diffident smile. We only have a day together, and although we will mention our adult children and compare the wildlife at his country home in Virginia to my country home in New Mexico and note that we are both college teachers–mostly we talk tiger beetles. The undergraduates from the 1970s crowd the house (Peter and I stay with a friend), some retired, some waiting to retire, but all projecting the spirit of the inner nineteen-year-old, devoted to a large cooler of beer and food, enthusiastic, but with pills to take now for high blood pressure and the wrong kind of cholesterol. They're excited to be here beside their beloved teacher. They're rubbing against each other, old intimacies and old frictions. They're snapping pictures of everything, the flowers in the yard, the birds in the trees, the ants and lizards. They're taking down the guidebooks from my shelf–Peterson's *A Field Guide to Western Butterflies, A Field Guide to Mammals of North America, A Field Guide to Advanced Birding.* There's so much to discover, so much to know! It would be daunting if we weren't all young and energetic, still waiting for life to reveal itself like a summer night, the stars appearing, one, twelve, fifty, and then the Milky Way, that nightly extravagance. The undergraduates from the 1970s look forward to seeing the Milky Way. Several hundred billion stars, fourteen billion years old.

Our own galaxy! We see it edge on, says the *National Audubon Society Field Guide to the Night Sky,* "like looking at a plate from the side."

Some of this happy crew join Barry and me this morning at the microscope at my neighbor's house. I have vials of dead beetles in hand, a humming in my heart: Barry will show me what to do.

And he does. We look at tiger beetles I had collected the day before and Barry explains that while the setae pads on the male's front legs are the best indicator of sex, a male's anterior is also more pointed, the female more rounded. Almost immediately, removing a female's dorsum, he abandons the tiny scissors as overkill and instead scrapes gently along the insect's side with one of the long thin pins I use to anchor the insects into paraffin. A few gentle scrapes, a tug, and the body cavity is exposed. Barry points out the ovaries, with three or four small eggs and one larger egg. We do another female, from earlier in the summer, to look at an undeveloped ovary.

Yes, Barry thinks collecting ten beetles every week is enough. Yes, storing them in 91 percent alcohol in these vials is working fine. Yes, it's great that I'm observing their seasonality–when they appear, when they aggregate, when they disperse. Yes, Barry says encouragingly, and I can see why his former students want to travel with him and why he has served me so well as a mentor.

Barry Knisley, Sharman Russell, and Becky Heck (photo by Dan Wheeler)

We drive to Bill Evans Lake and the Gila River Bird Refuge and Mogollon Box Canyon. With the recent flooding, the river is newly black with ash, and the Western red-bellied tiger beetles have left the water's edge, no longer in groups of ten to twenty per square foot, but two here on the path to the river, a group of three, another group of three. Skitter, dart, that movement of air. Armed with our guidebooks, we are looking at everything else as well, the first day of Creation in the Southwest—first a collared lizard with turquoise throat and belly, yellow face and hands, back and side mottled with cream and blue and orange spots, so that one would think "what a clown" if this reptile with a hard bite didn't look so regal. What a beautiful animal, Barry's former students murmur. And what beautiful plants. Rabbitbrush. Mesquite. Yucca. Cottonwood. Well, yes, I nod proprietarily. I think so, too.

Barry is particularly interested in Dick's observation of a pygmy tiger beetle in New Mexico. I call Dick and ask if we can come look at the very spot where he saw that pygmy. Twenty minutes later, Dick greets us at his front gate and takes us to a humble piece of American Lawn, USA Anywhere, patches of grass beneath some non-native tree watered faithfully through the summer. "This is it," Dick says.

We look at the ground. "What's that?" I ask.

Dick rushes to get a collecting vial, and there's the comedy of one world expert entomologist, one amateur entomologist (that's Dick), and one amateur amateur trying to catch a very small less-than-a-quarter-inch black tiger beetle. Barry spots a few more pygmies. This is a healthy population, but we are wearied with getting the one specimen—and that's all Barry needs anyway. Official confirmation. Dick and I grin at each other. Are we *good* or what?

July 23

Barry and his group leave in the morning. It's been a year since I saw that swarm of Western red-bellied tiger beetles eat a dead frog, slicing and scything, turning to me next as if daring me to enter their world. That primal ferocity still impresses. The profusion, the vitality. *You could spend a week studying some obscure insect and you would then know more than anyone else on the planet.* Of course, the Head Entomologist at the London Natural History Museum meant you could do this if you were already an entomologist, if you already knew your labial pulp from your maxillary pulp, if you were already handy around a microscope and fast on your feet with a collecting net. In my case, too, it's been a year, not a week.

But I've made progress.

Also today, I get this e-mail from Citizen Science Central at the Cornell Lab of Ornithology:

> From: Citizen Science Central
> Sent: Mon, July 23, 2012 11:31:18 AM
> Subject: have a minute? submit fake data!
> Hello! If this message reaches you by 2:30 pm eastern on Monday, would you take a moment to fake some data? Really! Sam Droege and colleagues at Patuxent Wildlife Research Center are beta testing phone, text, and e-mail formats for submitting cricket and katydid survey data. Later in the fall they will be hosting 2 events (Maryland and Hawaii) to survey calling orthopterans and in the case of Hawaii ... invasive frogs and geckos (the main idea is to gather baseline data, try out the technique and look for associations with the landscapes). You can help them do a real "trial by fire" by making a phone call, e-mail, or tweet now so they can find any weak spots before they go live. The fake event will be July 23 at 2:30 PM Eastern time and the "survey window" will be open for half an hour.

The chance to fake scientific data. Who could resist?

July 26

I pin the female tiger beetle into paraffin and deftly remove the elytra, or wing covers. A few scrapes of a pin along the side of her dorsum, a gentle tug, and the body cavity is exposed. The eggs look like tennis balls in panty hose. I count them. I move on to the next tiger beetle.

July 28

In the Gila National Forest north of Silver City, Cherry Creek Road is a trail more than a road, along the bottom of a canyon with an ephemeral flow of water over slick rock, not far from where I went with Mike to see

Western two-tailed swallowtail (photo by Elroy Limmer)

the Mexican spotted owls. Locust and wild rose flower along the edge of Cherry Creek, where I come here every summer for the butterflies: California sisters, red-spotted purples, hairstreaks, duskywings, skippers, and the Western two-tailed tiger swallowtail, three inches across, striped black and yellow, with spots of red and blue along the edge, and those improbable tail fillips. This is my favorite butterfly. I like its size. I like its design. I like how the males patrol canyons up and down looking for mates and carrion juices, smelling with their feet, seeing with simple eyes on their genitalia. The mix-up of senses is entertaining, but really it's that beauty passing by, the lazy lift of large yellow wings and glide of grace through interstices of pine. It makes my chest ache. I feel a movement in my ribcage—a lifting, a hollowing. I feel a yearning, whenever I see a Western two-tailed tiger swallowtail, that often shifts to happiness.

I'm walking here with my friend Marion and her husband Jamie, who has a new camera and who stops often to take pictures. Jamie is in love with the way he can grab pieces of the world, a macro shot of a blue lupine, the head of a fly, the shutter going in and out, whirring, getting closer.

Intimacy without violation. The camera drives him forward. Look at this! Look at me! The world clamors for his attention, his knowledge of light and frame and shutter speed. This is a new way of being alert, marveling, capturing–and then doing it all again, the world so full of itself and beckoning. This is a much slower walk with Jamie and his camera.

Marion stops and points to a movement in an oak tree, and I'm thinking, "bird" until I see the long white nose and red-brown fur and dark eyes of a coati staring intently from about thirty feet up. More movement in the tree and the coati climbs purposefully another ten feet and then back down, holding a kit by the scruff of the neck. Jamie is taking pictures even as we back away to watch the mother go down the tree and behind a bush, emerging again, then back up the tree to get a second kit, who is crying a little. Later this afternoon, we will load Jamie's photos into my computer and see the endearing nose of a third kit emerging from oak leaves.

In some areas–in tropical Guatemala–coatis make nests in trees and coati mothers also move babies whenever they feel a nest has become unsuitable. Perhaps that is what is happening now, the kits unsafe so high up, too mobile as they get older. But why would this coati do that in front of

Coati (photo by Elroy Limmer)

us? This seems to refute Chris Hass's idea that mothers deliberately lead potential predators away from their dens. Maybe, I theorize, this is a new mom, unsure what to do, where to put her rambunctious children, how to deal with these strangers on the road. Or maybe this is an experienced mom who knows that humans in New Mexico don't prey on baby coatis. Although she does not seem alarmed, we feel she should be, and so we quietly leave, hushed with the enormity of this gift: the way she went about the business of her day.

July 30

I dissect. I take notes. I feel pleasure in my tools, the dissection kit in its black wrapper, the instruments so small and precise.

Cinnamon black bear (photo by Elroy Limmer)

August 2012

August 4

The first large-scale open conference on citizen science–or as the organizers call it, Public Participation in Scientific Research–attracts some three hundred people to Portland, Oregon, a kind of Harry Potter city with its magical light rail transportation system, rose gardens, independent bookstores, good restaurants, bike trails, river walk, state-of-the-art recycling bins, and population of tattooed youth in training to be wizards. Portland is already rich with citizen science, home to the seminal Xerces Society, whose longtime mission is to conserve invertebrates and their habitats, with dozens of citizen science projects monitoring and protecting butterflies, dragonflies, beetles, crabs, mussels, starfish, and worms. Willamette Riverkeeper is also in Portland, organizing volunteers who collect and report water quality data up and down the Willamette River and its tributaries. Nature's Notebook is active here, and many other national programs. Almost everywhere you look–I imagine–public participants in scientific research are walking this city's parks and trails, counting birds, talking to dragonflies, making notes.

By 7:30 a.m., I'm out of the hotel and into the Convention Center, squirrel metaphors entirely appropriate. Next to registration is an enormous table of Danish, cinnamon buns, turnovers, bear claws, and muffins, backed by a row of thermoses with enough coffee to fill 1,218 bathtubs. The plenary speakers are beginning, and I move into a cave-like room of people settling into chairs, eating pastry, and listening to a talk on the early days of citizen science. In the United States, the oldest records were begun by farmers looking at crops, with our most important long-term observations made by volunteers working with the US National Weather Service. Much of what scientists know today about regional climate patterns, variability, and trends is the result of this cooperative network, as well as efforts dating back to the early nineteenth century. Later in that century, amateur naturalists like Henry David Thoreau started recording the leaf-out dates, flowering dates, and arrival of migratory birds in the eastern states. As a result, we can conclude that plant species are responding more strongly than birds to global warming.

The speaker segues smoothly: global warming–big changes across big landscapes–is one reason why more citizen scientists are needed today. Other large-scale problems like invasive species and emergent diseases also require massive teams and group effort. At the same time, citizen science is uniquely good at looking at small-scale research questions, from the decline of shark species near a specific coral reef to something interesting half a mile down your road.

Throughout the morning, other speakers point to the successes of citizen science. The glamorous Foldit protein-synthesizing program dreams of using its crowdsourcing video-gamers to create a molecule that will reduce the cost of producing hydrogen from water. The result could be clean, nearly unlimited energy made cheaply at power plants or at home with small plug-in hydrogen makers. In short, utopia. The equally glamorous Galaxy Zoo is comparing images from some of the oldest galaxies in the universe with some of the youngest. In short, answers to the beginning of time.

There's some discussion, too, of the main challenges faced by citizen science: 1) ensuring that the data collected is good data, 2) dealing with the enormity of data collected, and 3) sustaining the leadership and infrastructure of existing programs.

Like most scientific conferences, this one has poster sessions with sixty posters put up and down at specific times. A poster session is not a euphemism for anything else. Sheets of poster board with text and images about current research projects are hung side by side on six-foot-high panels that form rows, which people can walk through. Commonly, graduate students use posters to expand their resumes. They also stand in front of their posters and network. It's not all about science. Up and down the rows, I hear murmurs: "There's a job opening at–"; "He's hard to work with but–"; and "I'm applying for–". Most posters tout the value of citizen science, although one study highlights low accuracy rates in volunteers monitoring alpine flowering times in the Appalachian Mountains. A woman behind me notes that a similar study matched volunteer observations with those of botanists–and found significant errors in the experts as well. This starts a discussion between the woman and the PhD student and pretty soon they are exchanging business cards.

By now, I realize I am one of the few actual citizen scientists here. Mostly the three hundred attendees and speakers are scientists and program directors and professionals working in the field of citizen science.

Darlene Cavalier, former cheerleader and founder of the website Sci-Starter, confirms this as we sit together on a couch in one of the convention center hallways. I've arranged this conversation by e-mail, finding time for a coffee in her busy schedule. A very pregnant graduate student stops to say hello to Darlene, and we start talking children. Darlene's youngest of four is three and her oldest fourteen. This is the graduate student's first child. I mention how much I mourn those days of mothering–but how can either of these women understand? When you live at the center of the world, anchored as deep as a bay-spanning bridge, busy as the humming hive, you think you will always be needed in this way.

We talk a bit about citizen science, too. Darlene has new enthusiasms she is pleased to share: an upcoming Baby Laughter Project and the World Gratitude Map, which informally explores "the concept of resilience, the science of bouncing back from adversity, and how an 'attitude of gratitude' can make us happier and healthier." I am reminded of how much citizen science can explore the edges, less constrained by scientific convention, less afraid to look "unscientific."

Later in the day I magic across town via light rail to a dinner of Thai food with Jake Weltzin, director of Nature's Notebook and an ecologist with the US Geological Survey. He begins with the problem of organizing data. As one speaker has already noted: free data is like free puppies. Pretty soon your yard and house are overflowing with the gift of free puppies and the health inspector is knocking at the door. Jake admits that information about the phenology of plants and animals is scattered across too many programs. Project Budburst, co-managed by the Chicago Botanic Garden and the National Ecological Observation Network (NEON), also monitors the phenology of plants in North America. The online checklist for birds, eBird, was started in 2002 as a joint program of the National Audubon Society and the Cornell Lab of Ornithology; enormously successful, eBird has just reached its hundred million mark in observations. Journey North is a global study of wildlife migration and seasonal changes sponsored by the Annenberg Foundation. These and other smaller projects need to be coordinated into one database with one standardized protocol. Jake has been given that task, which could take years and will certainly require funding and diplomacy.

On a larger scale, the group DataONE (Data Observation Network for Earth) was recently launched by the National Science Foundation to establish a system for scientists and citizen scientists around the world to store, share, and access data on the Internet. Another online program called CitSci.org helps new projects organize by providing software programs to manage members and research, with a central website where the projects can network.

After dinner, Jake and I walk through downtown Portland, angling for the next light rail stop, sated with pumpkin curry, passing the block-long Powell's City of Books, enjoying the art galleries and parks and futuristic recycling bins. Jake and his family live in Tucson but once tried to buy land in the Gila Valley–the very land next to my backyard and Nature's Notebook site–because Jake was once the PhD student of a neighbor on my road who taught at the University of Arizona but who now lives off the grid and blogs about Peak Oil and climate change. My neighbor thinks we're doomed. The planet is simmering on the stove, and there's no hope for industrial civilization, for cities like Portland, for me, for him (although *he's* been caching away food), for my adult children. The apocalypse is nigh. Another neighbor, on my opposite side, thinks the same thing but for different, religious reasons.

Jake and I talk apocalypse, too. Who doesn't? Accumulating data about climate change is clearly important, but that's not the same thing as useful. I ask Jake how we can move from understanding climate change to mitigating climate change. How can citizen science affect national and international policy, the meta-systems of law and commerce?

Jake is a happy-looking guy who now puts on his serious face: yes, there's that. Projects with government support have some advantage. Budburst is tied to the National Ecological Observation Network, funded by the National Science Foundation. Nature's Notebook is part of the US Geological Survey. These connections help with the problem of sustained funding and assume a certain transfer of knowledge and influence. But ultimately, government programs rely on public support. So it comes back to the phrase *citizen scientist,* in which the two nouns are given equal weight. Citizen of a community, citizen of a country, citizen of the world–the ball is in our court. Going even further, while science is nice, engaged citizenry is the transformative element.

August 5

In two days of speeches, coffee, and pastry, what surprises me most is the growing role of citizen science in environmental activism. Ben Duncan, a policy analyst at Oregon's Multnomah County Health Department, talks about bucket brigades in which community volunteers measure air quality using a plastic bucket with a simple pump system; these air samples are then sent to labs and tested for gases and sulfur compounds. The Louisiana Bucket Brigade, one of many such projects around the world, has a ten-year record of documenting toxic emissions in neighborhoods near oil refineries and chemical plants. Their team of citizen scientists begins complaints to officials with, "I know you say that billowing black smoke outside my window is harmless, but I'm getting a high reading of benzene that violates state standards."

I meet for coffee (more coffee!) with Muki Haklay, professor of geographic information science at the University College London and co-founder of ExCiteS (Extreme Citizen Science). Muki came to Portland to promote what he calls "the logical next step." Many of the urgent problems in pollution and habitat destruction occur in places where people are poor and illiterate; extreme citizen science wants to empower these communities to address research questions important to them. In the Congo Basin, for example, pygmy hunters and gatherers are interested in documenting deforestation and illegal poaching: where the poaching camps are, what animals are being hunted, the condition of the forest. They use a cheap, durable, waterproof smartphone with GPS, on which they record their data, which is sent and coordinated into a larger map. Importantly, the phone was adapted for people who don't read, using images instead of words, and is powered by a device that generates electricity by heating water over a campfire. Scientists at ExCiteS help with the mapping and analysis. Muki tells me, "Our ultimate goal is to develop better tools that allow communities to view and analyze the data themselves. We want to take ourselves out of the equation."

Wallace J. Nichols, a herpetologist at the California Academy of Sciences, begins his talk with the anecdote in which a scientist and tribal elder swap creation stories. The elder explains that the world sits on the back of an elephant who stands on the back of a turtle. The scientist wonders what the turtle stands on (a turtle) and then what that turtle stands on (a turtle) and what that turtle stands on (a turtle) until the elder sighs, "No more questions, young lady. It's turtles all the way down." In the early 1990s, Nichols's fascination with turtles took him as a graduate student to Baja California, where the hunting of sea turtles had just been declared illegal. Overnight, fisherman who had caught and eaten turtles all their lives were criminalized, even as a new black market for sea turtles flourished. At that point, scientists considered the five species of sea turtles native to the Baja coast a lost cause.

But Nichols listened to what local men and women had to say and disagreed. Over the next fifteen years, he and others worked with the people living along the Sea of Cortez. They formed the conservation organization Grupo Tortuguero. They helped protect nesting sites. They educated the public about pollution, habitat loss, and the overharvesting of turtle eggs and turtles for meat. They discovered that loggerhead sea turtles travel back and forth between Japan and Baja California to reproduce. They began to write scientific papers based on their research.

Nichols emphasizes that all this required trust and transparency. Scientists and citizens were equally engaged in conservation and research, which is now done mostly by nonscientists living in the area, as well as volunteers from environmental and citizen science groups. The research itself, Wallace J. Nichols tells us, is "pretty damn good" with papers in a number of peer-reviewed journals. He ends his talk by comparing his work with citizen scientists to some starfish species who reproduce by dividing or breaking in two, creating new starfish, who go on to divide into starfish themselves, who go on to create more starfish.

Afterward, one starfish tells Nichols his story. As a college student, he thought it would be fun to work with Grupo Tortuguero for a summer, a decision he briefly regretted when his boat got caught in a storm and he was stranded on an island without food for two days. "It was so great, so awesome," the still-young man says now, pumping Nichol's hand. "I went on to get a masters in science education and am now working on my second masters in museum studies."

An admiring group has formed around the charismatic and handsome Nichols, who explains that it is really "turtles all the way up" with one awesome experience in citizen science leading to another awesome experience–and with the population of sea turtles along the coast of Baja California increasing exponentially. In the 1990s, green turtles there numbered five hundred; now there are five thousand. All this is good for conservationists and scientists and tourists and fishermen, too.

The point is not how citizens can serve science–but the other way around.

August 13

While I was in Portland, Peter collected beetles for me again, slapping his net in the grass along the Gila River and catching seven (he declares) at a time, like the brave little tailor in the Grimm's fairytale who killed seven flies in one blow. The tailor made himself a belt with that motto, Seven in One Blow, and went off on adventures that led to marriage with the king's daughter. "Do you want a belt?" I ask Peter. He declines.

Today I am at Mangus Creek, near Bill Evans Lake, where Western red-bellied tiger beetles are plentiful–although last week, when the creek was dry, they were entirely absent. Clearly these insects do not simply disperse but come and go in response to habitat change.

On my Nature's Notebook walk to the river, I see that motion of air deliciously familiar. This time it is a punctured tiger beetle, commonly found through much of the United States and southern Canada. In the summer, punctured tiger beetles can also be seen on city sidewalks and parking lots. Without any markings, this species is easily confused with the black sky tiger beetle. (You have to stop and admire that name, just as you have to wonder–why punctured?) However, "In the hand," my guidebook says of the punctured tiger beetle, "the labrum has only a single middle tooth." My punctured tiger beetle is a dull olive-brown, although variations within the species can be bright metallic green, which naturally I would prefer. I add the punctured to the summer's pygmy and ocellated and Western red-bellied: four species altogether. Adding another species to my list makes me feel oddly energized, like a collector with a new ceramic pig or a wannabe-retiree totting up her bank account or maybe just a kid singing a song: *I saw a punctured tiger beetle, I saw a punctured tiger beetle.*

August 18

Peter and I and my friend Shirley are driving to the alpine community of Nutrioso, Arizona, to take a tracking workshop. For nine years, the leader of this workshop developed and directed the citizen science program Wildlife Linkages, volunteer trackers who document wildlife sign in Arizona for the conservation group Sky Island Alliance. Their accumulated data–fifty transects monitored and one thousand track surveys–has helped win approval for two wildlife underpasses and one overpass across highways near Tucson. The organization regularly consults with the Arizona Department of Transportation, hoping to create a world in which departments of transportation regularly consult with bobcats and bears.

We start with the difference between felines and canines. Both have four toes but cats–domestic, bobcat, mountain lion–have an obvious leading toe, their claws rarely showing, with the three lobes at the bottom of the pad usually distinct. The overall print is round. Typically, the claws of a canine–dog, coyote, fox, or wolf–do show and the three lobes of the pad may be blurred into a single line. Often a dog track has a mound of dirt in the center. The overall print is rectangular, and you can draw an X through the print without crossing through any of the pads.

A bear print, with five toes, looks like a barefoot human footprint except that the big toe is on the outside. A coati also has five toes and, like a raccoon, makes a track that resembles a human handprint, but with the pinpricks of claws.

Then there are back legs and front legs and how fast the animal is walking and soil surface and erosion and multiple tracks and the way the light is falling on the tracks and a dozen other circumstances and vagaries. As with most new skills, what increases first is my level of uncertainty. I learn to say with confidence, "I don't have enough knowledge or information here to tell you who made this track."

We go out in the Apache-Sitgreaves National Forest, which surrounds the house where the workshop is taking place. We don't see many tracks today except for the fake ones made by the workshop leader, who walks

 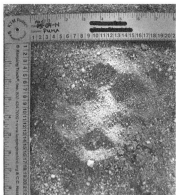

Left: Bear track; right: mountain lion track (photos by Sonnie Sussilo)

ahead with a bag of plaster molds. Oh, gosh, a bear. But I can see the bear print she holds behind her back. We are looking not only down at the ground but up and around at the way this forest is recovering from the Wallow Fire last year, which burned 538,000 acres in Arizona. Nutrioso and nearby communities were evacuated, and it's still a wonder that our host's house, in the middle of the forest, didn't burn.

I can hardly see the blackened trunks of ponderosa pine for the wildflowers beneath them. Red skyrocket and blue lupine and white yarrow and purple aster. Indian paintbrush and salsify and daisies and groundsel and goldenrod and feathery grasses. Tall mullein. Tiny dayflowers. It's like walking through a bouquet. The visible metaphor of joy. A celebration. A party! Too many wildflowers and I'm overstimulated, like a hyperactive child.

Some of the larger burnt pines are alive and will recover, although many more will not. Low-intensity and periodic fire may be good for fire-adapted ponderosa-pine forests, but last year's Wallow Fire in eastern Arizona burned too hot to be part of that cyclical process. Fire suppression over the last hundred years and the resulting build-up of underbrush and crowded trees, as well as clear-cut logging, "sort of broke the structure of the forest," said fire ecologist Stephen J. Pyne in a newspaper report after the fire. Large swathes of this area are still a sea of black sticks, while habitat for species like the Mexican spotted owl disappeared overnight. On Escudilla Mountain, a few miles from this tracking workshop, all the spruce and fir went "poof."

Escudilla Mountain is a name that reverberates, the place in 1914 where the young hunter Aldo Leopold gleefully fired into a family of wolves, a "welcoming melee of wagging tails and playful maulings," killing one old female and reaching her "in time to watch a fierce green fire dying in her eyes." Years later, Leopold wrote his famous essay, "Thinking Like a Mountain," in which he realized that ecologies require both predators and prey, wolves and deer, and that too many deer meant too much browsing and a denuded landscape. This, at least, might be the mountain's train of thought. The epiphany helped shape Leopold's idea of a land ethic and became part of America's conservation movement.

In my mind, too, Escudilla remains a thinking mountain, imbued with its own humanity, all those college students still reading Aldo Leopold, all of us who walk its trails. From one perspective, the recent conflagration simply hurried the mountain's natural process. Typically, mature spruce-fir forests die by being consumed in crown fires, with aspen growing up in their place, providing shade and habitat for the next generation of spruce and fir. Aspen are growing up now after the Wallow Fire, on Escudilla Mountain where Leopold helped rid the Southwest of its native wolf, and where wolves are slowly and painfully being reintroduced. But in this case, with the looming specter of global warming, the spruce and fir will not likely return. That story is over.

That story is probably over as well in the Gila National Forest, where the Whitewater-Baldy Fire crowned and exploded through large patches of spruce-fir in elevations above eighty-five hundred feet. Those forests are also gone now, and aspen are also growing in their place, by some reports shooting up five, six feet since the summer rains started. In other ways, the recent Whitewater-Baldy Fire was less destructive than most Western blazes because of a previous program of prescribed burning and allowing some wildfires to burn naturally. Even so, continued drought as well as pine beetle outbreaks still threaten the Gila's remaining stands of pon-derosa pine. In a reasonable–and possibly best-case scenario–those trees will eventually be replaced by more heat and drought-tolerant species like oak and juniper. A worst-case scenario includes repeated fires that prevent this transition, resulting in ... what? A desert of mesquite and shrub? A landscape we can't yet imagine?

Almost any change will be painful for humans like myself, living here for our span of decades, the Mexican spotted owls gone, the columbine gone, the absence in the air of Western tiger swallowtails. Even the Gila River could end up drying, the cottonwoods and sycamores yellowing, falling, limbs tangled, bleached bodies in the sun.

The ululation begins.

But none of that is happening now. Now, with a little more training, I could join Sky Island Alliance's first volunteer group in New Mexico in the Burro Mountains twenty miles from my home in the Gila Valley. Every six weeks, this team walks a mile of a designated streambed and records any sign of keystone predators–black bear, bobcat, mountain lion, white-nosed coati–and any threatened or endangered species such as Mexican gray wolf, jaguar, or ornate box turtle. They measure and photograph all tracks, scats, scrapes, and kill areas to send to Sky Island Alliance, which puts the information into a national database.

The tracking team's first visit this August was spectacular, as a friend of mine reported in an e-mail. "It was a palm the size of a man's hand. Black bear. Lots and lots of black bear. We were on the I-10 (or I-95 if you're east of here) of bear highways." The bear tracks became so numerous that the team got bored with them. Just as they were saying for the third time, "We should turn back," they saw the print of a different animal. "Four toes. No X in the middle. No claw marks. Longer second toe. Mountain lion. Big guy just strolling along down the middle of a game trail a little above the streambed."

A mountain lion makes sense. A highway for bears would also be a highway for lions and foxes and coyotes wanting to get somewhere quickly. A good place for deer and javelinas, too, which explains the lion strolling along just a little *above* the streambed. Lions need eight to ten pounds of meat a day and will kill a deer every nine to fourteen days. All this requires dedicated strolling.

Peter, Shirley, and I leave the tracking workshop, driving home through patches of burned and unburned forest. We talk about tracks. We talk about mountain lions. We haven't heard of any more lions being killed in the Gila Valley, perhaps because people are penning their livestock more securely or perhaps because the lions who live here are more wary of human presence, winding more secretly through bushes and scrub brush, using the river as a corridor as they always have used the river as a corridor, as they use culverts in cities like Tucson and San Diego, as they move silently through the suburbs of Silver City and other towns, lacing wildness into our lives, their story still part of our own.

Four toes. No X in the middle. No claw marks. Longer second toe.

That's what's happening now.

August 24

The Western red-bellied tiger beetles have largely and finally dispersed. Even the shoreline around Bill Evans Lake is empty except for a handful of skitter and dart. I collect my ten beetles, but slowly, going to three different sites. Later in the day I visit with Allison, the woman from the How to Become a Leading World Authority Club who first introduced me to Nature's Notebook. We compare literal notes, my mistake confusing a catclaw acacia with an Eastern whitethorn acacia, and hers misidentifying an alligator juniper for a one-seed juniper. "Now I know the difference," she says.

More to the point, in preparation for the fire threat this summer and with funding from the Forest Service, tree-thinning professionals took out eighty trees on her seventeen acres of land with its eleven tagged plants. The bulldozers drove over the blue and side-oats grama grasses. The ponderosa and piñon pine are gone. Other trees on her property are dying because of bark beetle infestation. Allison grows silent, contemplating the wreckage in her backyard before continuing, "But I'm going to start again next spring!"

She asks about Ed Greenberg's project of growing leaves hydroponically for their micro-nutrients. I tell her that Ed seems to have dropped the idea for now. When I saw him recently at the hardware store, he talked mainly about the outside oven or adobe *horno* he had just built for Silver City's new food pantry and Center for Food Security and Sustainability. I'm on the board of the organization that runs the food pantry, and Allison and her husband are on the tracking team with Sky Island Alliance, and we can now point to any number of mutual friends and connections. That's because this is a small town and the connections easily seen; we hardly even marvel at them anymore: the crisscrossing of overlapping, entangled threads creating a pattern of light, sometimes gossamer, sometimes brilliant, a light we live in whether we notice it or not.

August 25

Yesterday the Gila River Bird Refuge at the end of the dirt road was so eroded by a recent rainstorm that I had to turn back. Today I park the car where the road becomes a hole and walk two miles to the river, where I don't see any tiger beetles at all. On my return to the car, I pass the hackberry trees and look for coatis, thinking off and on about two people who are premier citizen scientists here in southern New Mexico. One is an emergency room physician who builds and operates telescopes accurate enough to send useful information to NASA. The other retired from being a surgeon (mine was the last colonoscopy he performed) and became an expert on the bryophytes (mosses, hornworts, liverworts) of the Gila National Forest. I've been talking with both men lately and am struck that neither felt transformed by their work as citizen scientists. Instead they built on known skills, pursued longtime interests, and kept using the same sides of their brain.

My goal, pursuing the Western red-bellied tiger beetle, was more grandiose. I didn't want to be something else so much as someone else. I wanted my experience as a citizen scientist to transform me, and I wonder if this is a personal theme. Moreover, is it wrong-headed? The desire to transform implies that whoever you are to begin with is not good enough; self-doubt doesn't seem a strong starting point for change.

At the same time, some citizen scientists *do* uncover hidden strengths, neglected strengths, and that's surely a good feeling. They find themselves surprisingly adept at folding amino acids or cataloguing galaxies, usefully studying urban squirrels or phytoplankton. They are simpatico with clouds. They are good at making charts.

Where an arroyo meets the dirt road, I stop and look for tracks. A few feet up the streambed are a nice set of bobcat prints. There's no mistaking that roundness, the leading toe, the size of the front and back feet. I also see a fox print, or maybe a small coyote. Foxes are on my mind since I saw three earlier in the day, probably a mother and two kits, who ran so quickly into the brush I spent a few minutes questioning what I had seen. Was that a fox or a wish?

Bobcat (photo by Elroy Limmer)

That's one good thing about tracks. They stay there. You can admire them for long minutes, imagining the animal who passed by, feeling the tangible presence of a bobcat, short-tailed, tufted-ear, delicately-spotted, charismatic, predatory.

It's another gift, the world showering us with gifts, the tail of a fox, tracks in the sand, and there–growing up the shadowed bank of this arroyo, a mound of jimson weed, also called moon flower, also called thorn apple, also called sacred datura, the large, creamy, lavender-tinted, trumpet-shaped blossoms seeming to glow, exuding power and a rich scent.

And next to the flower, here in this streambed, a massive dark rock with white radiating lines, a geometric pattern of dark and light, veins of quartz, cool to the touch.

Is all this for me? I feel the need to fall in love with the world, to forge that relationship ever more strongly. But maybe I don't have to work so hard. I have thought nature indifferent to humans, to one more human, but maybe the reverse is true. Maybe the world is already in love, giving me these gifts all the time, calling out all the time: take this. And this. And this. Don't turn away.

Sacred datura (photo by Elroy Limmer)

And here's my car, too, at the end of the road, the end of this field season, my second summer chasing tiger beetles. My day pack clatters with glass vials for the beetles I didn't find, and once again, no different today, I didn't discover any hidden or neglected talent. I didn't transform into someone I am not. But I am building on existing strengths. I've always been good at skittering and darting on the surface of things, adding a few animal tracks to the few birds I know and the few butterflies and the few plants. And I've always been good at walking a country road and seeing for a nanosecond something shivery and grand. Requited love. I am the bride of the world, and I am the groom.

August 30

In the last month, I have been busy at the microscope, dissecting the ovaries of the tiger beetles I collected all summer and stored in vials. I have also sent female beetles to Barry Knisley for him to dissect. His results, e-mailed today, immediately show me how I should have been busier. I should have noted which ovary those eggs were on. I should have measured my eggs with a special device I didn't have. I should have first separated the males and females for each group, since the number of males was almost always higher, and for many groups I ended up dissecting all the females with none left to send to Barry. I should have practiced on other tiger beetles before I dissected the ones I collected. But now I know. In any case, our separate investigations more or less dovetail and that is a miracle outweighing any self-reproach.

I write up a report collating our results.

For the vials marked June 10, my observations were that every tiger beetle I collected was male. In a series of rapid-fire e-mails back and forth throughout the day, Barry explains, "The predominance of males in collecting is often a problem, and I just had the same experience with *C. floridana* this week. Often I try for the ones that look bigger. It may be males are more common, but usually females seem to be less approachable or fly farther so males simply get collected more often." For June, also, it could be that Western red-bellied males were emerging earlier than females, jockeying to be the first to find and fertilize a female. This would be true, at least, if the females were ready to be fertilized and lay their eggs soon, since it is not the first male that wins the prize but the last one to eject his sperm.

For the vials marked June 17, my observations were three males and two females with undeveloped ovaries. This was my first dissection after Barry had shown me how, and my notes are mainly concerned with sexing the beetles. "Very clear setae bristles," I wrote. And "Maybe no setae but male genitalia?" And "What am I looking at here?"

For the vials marked June 30, my observations were five males and three females, each female with cheesy-white ovaries, no eggs or "maybe eggs?" Barry's observations were one female with distinct but undeveloped ovaries and no eggs evident, and one female with three small, immature white eggs (.5 millimeters) on the left ovary and three medium-sized white eggs (1 millimeter) on the right ovary.

For the vials marked July 6, my observations were five males, one female with cheesy-white undeveloped ovaries and no eggs, and one female with many small white eggs. Barry's observations were one female with undeveloped ovaries and no eggs, and one female with five large white eggs (1.5 millimeters) on the left side and four large white eggs (1.3 millimeters) on the right side.

Stop right here. A month from their appearance on the Gila River, some Western red-bellied tiger beetles already seem ready to oviposit, since large eggs are an indication of that. But others have only small eggs or are without eggs in undeveloped ovaries. Of course, we don't really know much about egg development. Females may retain large eggs for some reason—perhaps they need more maturation, although Barry has never seen that in the lab, where small eggs progressively develop to large and then are oviposited.

However, our hypothesis had been that these insects would only have mature eggs late in the season, in August, just before they disperse from the water's edge to oviposit in upland areas. It looks like that hypothesis was wrong. Perhaps, instead, tiger beetles are dispersing throughout the summer. Once they mate and feed and develop mature eggs, they leave. But then wouldn't I have seen a greater and steadier decline in numbers? What I saw instead by the Gila River and Mangus Creek (what I think I saw since I didn't do a systematic count) was a temporary dispersal when the habitat flooded and then a return of beetles, which remained abundant into mid-August.

For the vials marked July 14, my observations were eight males, one female with undeveloped ovaries, one female with several small white eggs, and one female with several large white eggs. The trend is continuing: a mix of stages.

For the vials marked July 20, my observations were eight males, one female with five or more middle-sized and large white eggs, and one female also with five or more middle-sized and large white eggs.

For the vials marked July 29, my observations were five males, one female with three or more large white eggs, and one female with five or more small white eggs.

For the vials marked August 4, my observations were six males, one female with four or more large white eggs, and one female with three or more large yellow-orange eggs. Barry's observations were one female with three large (1.2 millimeters) yellow-orange eggs, plus additional smaller white eggs on the right side and two large yellow-orange eggs (1.2 millimeters) and additional smaller white eggs on the left side.

The color of the eggs is new and exciting. Barry wonders if this is a sign of immediate pre-oviposition. But there is nothing about this in any of the literature on tiger beetles. Later in the day, David Pearson e-mails his thoughts, "We know that the number of eggs and perhaps mean egg size varies with the amount of food the mother received while developing the eggs, but I have heard of no such evidence for egg color variation in tiger beetles. Some butterfly eggs change in color and density as they approach hatching. That may be what is happening here. A far-out hypothesis is that there is some evidence that a few frogs and insects lay some eggs that are available only for their other young to eat. They are usually larger and of a different color, but again I know of no evidence for this among tiger beetles."

For the vials marked August 12, my observations were seven males, one female with undeveloped ovaries, and one female with many large white eggs of a somewhat rectangular shape. My note about the shape is probably meaningless. Still, I have become a better note taker, drawing little pictures. Barry's observations were one female with undeveloped ovaries.

For the vials marked August 20, my observations were three males, one female with one large white egg, and one female with undeveloped ovaries.

For the vials marked August 24, my observations were that all the beetles were males.

In the end, our limited study suggests a progressive development of eggs that starts reasonably soon after the beetles emerge—not at all as we had predicted. Eggs are in all different stages throughout the season. The orange color of the eggs in August remains puzzling.

✳ ✳ ✳

Seventeen years ago, I had a pertinent conversation with the archeologist Patty Jo Watson. I was writing a book about archaeology and Patty Jo was reminiscing about her early days in the field, looking at the origins of agriculture in North America. "Everyone thought then, in 1976," she told me, "that gourds and squash were introduced here from Mexico and that sunflower and sumpweed were indigenous domesticated plants. We were rooting for sunflower and sumpweed to have been cultivated first, before the introduction of gourds. We wanted eastern North America to be growing their own crops early on." When Patty Jo and her colleagues excavated a shell mound in Kentucky, they thought they would find some nice antecedents of modern plants, some slightly smaller sunflower and sumpweed seeds. But they didn't. "Instead," Patty Jo said, "there were several small charred fragments of *Cucurbita pepo* gourd rind. That was a wonderful moment for me. The gourds came first! Our working hypothesis was totally, unequivocally wrong. The archaeological record had spoken. It confirmed my faith in the science of archaeology."

Scientists would later date one of these gourds as forty-five hundred years old, too early to have been introduced from Mexico. The archeological record had more to say. Most archeologists now think that the entire complex of eastern American crops was indigenous, including gourds and squash, excluding late arrivals like maize and beans. The news has produced little fanfare, but eastern North America is one of the world's primary food production centers, a place where people independently developed agriculture.

What remains with me is Patty Jo's delight at being proven wrong. Science depends on such testable hypotheses. Our hypothesis that the eggs of the Western red-bellied tiger beetle would mature late was wrong, and learning that was another wonderful moment.

I consider the idea that females lay eggs along the edge of the Gila River despite regular flooding. After all, Eastern beach tiger beetle larvae can survive underwater for as long as six to twelve days (these larvae also form hoops, remember, and roll across the sand) and two Amazon species live underwater for three months. But the absence of larval holes along the river, the force of its flooding, and the fact that the females in my terrariums did not lay their eggs on sand makes me put aside that possibility.

Instead, Barry and I guess that individual females leave the water's edge to deposit eggs as the eggs are fertilized and matured. Then the females return to the river to continue feeding and mating. This would be similar to the aridland tiger beetle, which travels at night up to one-half mile from her feeding area to lay her eggs in nearby sand dunes.

What's crucial for all tiger beetles is to time their development so that adult males and females emerge together from the pupae and mating can occur easily. If Western red-bellied females lay eggs throughout the season, perhaps first-instar larvae hatch at different times—limiting competition for food—and then overwinter in different stages, first and second and third instar, with the third-instars all waiting for the same cue of temperature

Ocellated tiger beetle (photo by Giff Beaton)

or day length to pupate and come out as adults in early summer. Pupae of some species also overwinter, although this is not common.

Barry writes, "Continuing next year would be useful. Comparing ovary development with another species that we know oviposits 'normally' in the habitat where adults occur would be valuable. A series of collections and dissections of the two different species would provide additional insight." That species, I think, would be the ocellated, not plentiful in this area, but always present. Barry further wonders if yellow eggs would be seen in ocellated ovaries—and what would that mean?

"*C. sedecimpuncata*, your Western red-bellied, is such a widespread and super abundant species," Barry concludes. "In the Southwest it must be very successful in ovipositing many eggs with lots of larvae surviving."

To sum up—and personally I need this kind of repetition—at lower elevations in Arizona, Western red-bellied tiger beetle adults appear in early June and congregate around drying water sources where they can be seen feeding on helpless organisms such as tadpoles and small fish. They disperse after the rains have started and appear at lower densities in the uplands. Here in the Gila Valley, at a higher elevation, this species also appears in June and congregates at the bank of the Gila River and other water sources. They disperse if their habitat floods but return when the water stabilizes. Ovary dissections this summer revealed large eggs, seemingly ready for oviposition, a month after the adults' appearance, also after the rains had started, with eggs in all stages maturing throughout the summer. Oviposition, then, probably happens relatively near the habitat where the adults congregate and feed into late August.

In the Gila, this September and through October, I would now expect to find larval holes within half a mile of the riverbank–and preferably closer. I envision certain areas I have walked by many times, easy to find and dig into, peppered with almost-perfect circular holes. I suspect I will only find second or third-instar holes, since the first-instar holes are so very small. I imagine shouts of gladness.

And those orange eggs? A quick search on the Internet reveals that orange is a "very positive color," a power color, a healing color, mixing vitality with endurance. One blogger says that if a change is needed in your life, burn an orange candle for seven nights. Another website confirms that "With its enthusiasm for life, the color orange relates to adventure and risk-taking, inspiring physical confidence, competition, and independence. Those inspired by orange are always on the go!"

Western horse lubber grasshopper (photo by Elroy Limmer)

September 2012

September 4

All summer, there's been trouble in the terrariums. In June, the beetles I collected died and then again in July, for no obvious reason. I cleaned out the plastic boxes and filled them with new soil, and now I have seven pairs of mating males and females and four first instar holes, four baby larvae down in their tunnels, waiting for food. But I've had a difficult time getting mini-mealworms. Some companies won't ship live worms because of this season's extreme heat, and some companies don't have mini-mealworms because their supply died. Two batches of mini-mealworms I received by mail expired immediately in the refrigerator. Sometimes, instead, I use slivers of raw pork or chicken. I've also caught and dismembered a spider and a grasshopper. I'm about to try a small fresh muffin with organic blueberries.

September 8

The Sky Island Alliance team has invited me to help monitor their tracking transect in the nearby Burro Mountains, and we mosey along seeing signs of deer, javelina, cows, cows, cows. Maybe the footprint of a coyote, maybe a fox. The dry, chalky soil of the Southwest usually retains the prints of animals for days. But after a rain last night, the soil is damp, and tracks harder to see.

Here's a dead coati, dead for some time, at the top of the canyon.

Here's some scat. And here. And here. And here. And here.

I enjoy walking with Allison and her husband and the other team members, looking down at the ground. Everywhere I go now, I'm looking down at the ground, searching for larval burrow holes.

September 10

In the terrariums, males ride females for hours, with the few lone males circling for advantage. One male is strikingly aggressive, more than the other unattached males, more than any tiger beetle I've seen. He grabs another male on top a female. The second male is firmly lodged, but the attacker doesn't give up, scrabbling and biting until I'm alarmed at the intensity of the struggle. That insects have personality is an observable fact, although one we do not often notice. Certainly we see differences among species, with some butterflies demonstrating longer memories when we set them to certain tasks in controlled experiments–and within those experiments, some individuals doing slightly better than others, behaving differently based on physiology and life experience.

Other tiger beetles in the terrarium are throwing themselves on the ground and flailing their legs–my god! I think, they're having a seizure!–and then picking themselves up, calming themselves, so to speak, pretending that nothing strange just happened. Perhaps this is grooming behavior, rubbing off dirt or fungus, using the ground as a towel.

I have a Zen-like moment.

Not in the field, not in the beauty of sun-splashed river and dappled shadow and scented plants, but surrounded instead by green and orange plastic terrariums, cheap heat lamps made in China, my Obama/Biden '08 poster on the wall, and a shelf of books–all these inorganic angular lines. It is a moment in which I feel no separation. In which I am or imagine or believe myself all these things, molecules swirling, the dance of matter and energy in our separate forms, the play of forms: the Western red-bellied tiger beetle, the fungus in the terrarium, the book on the shelf, the metal lamp. We play at our need to mate. We play at the emission of heat and light. We stream through the window in packets of photons. We think about towels, children, approval. Out of the original void...how did such a moment come into existence? Why these colors, the privilege of being here? The quantum field. The beauty of physics. The physics of beauty. Only the human is asking such questions, her neurons firing in just this way, ten billion nerve cells linked by branching dendrites.

The moment passes. The tiger beetles stagger and stab, and I empty the terrariums of extra males, letting them loose in the backyard. All this drama is bad for my first instars.

September 15

My Nature's Notebook walk, those three acres behind my house and irrigation ditch, has been overtaken by an invasion of six-foot-high kochia, a drought-tolerant member of the *chenopodium* family brought to North America in the early 1900s from the steppes of southern Russia. Also known as poor man's alfalfa and sometimes grown as a low-cost feed, the red-stemmed plant with slender leaves is brown now and drying, seed-heads bursting, stalks sharp, a prickly mass tangled with wolfberry and hackberry and, of course, tumbleweed—another exotic from Russia, another thorn in the side of the West. Game trails snake through this sudden dense field, paths too low and narrow for me to use. Instead I have to shoulder through the kochia wall of thorns, snapping, flinching, accumulating debris. My socks grow needles. My clothes harbor enemies. Clothes are preferable to being naked, of course. What a thought.

At this time of year, I stop going on certain hikes precisely because of this bad plant behavior, the way these species spread their seeds in hardened, sculptured capsules designed to catch on surfaces and be transported elsewhere: the spiny cocklebur and showering pins of Bigelow's beggarticks, the common stork's bill with its augurs and upward-pointing barbs, the horns of devil's claw enclosing a shoe, the truly devilish goathead piercing right through that shoe's leather bottom—and sometimes into the tire of your car. Today I have no choice but to mince and barrel through these yielding waves of reproductive animosity, needing to check on my list of plants. The desert willow has over one hundred dry and brown seedpods with 75 percent still unopened. Unlike neighboring trees, my honey mesquites have lost their fruit, no long crunchy bean pods, with 90 percent of their leaves green. The leaves of the male four-wing saltbush are also still green, tiny bladders on those leaves storing and then releasing salt, which may act as an antifreeze for the cold months ahead.

It is the female saltbush that makes me pause and feel tender, that maternal figure almost hidden—transformed, burdened—under a massive cloak of brown, papery, four-winged seed. Native Americans once ground those seeds into flour; birds and small mammals eat them today. My neighbor's goat nibbles on them, too, with that intent goat expression, like someone munching potato chips.

Painted grasshopper (photo by Elroy Limmer)

Back at my house, the native sunflowers are growing up wild and well watered in the orchard and courtyard, so that they tower eight, nine, ten feet high, forming tunnels and bowers, topped by bright yellow flower faces. It's like living in a picture book. I expect a giant ladybug. A talking caterpillar.

I continue on to the river, my second Nature's Notebook walk highlighting grasshoppers. And sound. Rustle, thud, ping. Grasshoppers landing, jumping, falling against stalks of grass, on the dirt road, at the edges of the road, in the encroaching kochia. Their hops and crashes leave shallow holes in the dirt. Grasshopper tracks. Small brown grasshoppers everywhere, like stars in a summer sky, but if I looked closely I would also find the painted grasshopper in retro-sixties psychedelic, and the beautiful Western horse lubber grasshopper, black and yellow and lacy green.

I check my plants. The American plum has no new or discolored leaves, its fruit snapped long ago in a May freeze. The black willow also has no new or discolored leaves; the catkins of this male will come out again next early spring. Meanwhile, the rabbitbrush has thousands of small buds, 25 percent flowering.

The cottonwood tree stands isolated, near the field where sandhill cranes will come later to search for grubs and roots. Sixty feet high, with a deeply fissured trunk twenty feet around and extended limbs, the tree's leafed canopy covers an area as big as my house. Its triangular leaves attach to a long leaf stem designed to flutter so as to maximize exposure for the intake and outtake of carbon dioxide and oxygen. Last spring, from March to April, this female had flowers in drooping catkins; the fertilized fruit or seed attached to silky white hairs that float in the wind, cotton fluff, which looks alarmingly like an allergen but is not. (The male cottonwood pollen, however, rates a nine on the allergy scale of one to ten.)

Starting in the summer and clearly seen now, these branches have sticky leaf buds that can be made into an antibiotic, pain-relieving salve also known as the Balm of Gilead, a Biblical reference to a bush growing in the mountainous region of Gilead. Other parts of the cottonwood are also well-known remedies since the bark contains the precursors to aspirin, reducing fevers and inflammation. Every plant in my Nature's Notebook has some medicinal property. Willow, too, contains salicyclic acid, which relieves pain. The male blossoms of the four-wing saltbush make a wash for ant stings. The gum of honey mesquite is a disinfectant, an antacid, a treatment for lice. Rabbitbrush can be made into cough syrup. Yucca is a steroid.

Almost everywhere I walk along this dirt road, through that dense field of thorns, I find plants woven into the chemistry of my body. For menstrual cramp, take stork's bill. A poultice of Russian thistle is good for insect bites. Kochia may prevent metabolic disorders, from vascular disease to obesity. (Mice fed high-fat foods were given an extract of kochia seeds and didn't gain weight.) Cocklebur can be used for malaria, diseased kidneys, and tuberculosis. The invasive goathead has long been popular in Chinese medicine as a tonic; animal studies suggests it boosts sex drive. How can I doubt my place in the natural world, my sexuality linked to goathead, my kidneys to cocklebur?

Phytoremediation is the name for a new field of environmental science: *phyto* meaning plants and *remediation* the act of repairing or healing. The Environmental Protection Agency is now testing the ability of some plants to absorb toxic metals, which the stems and leaves then hold in their cells as a defense against insects and infection. Poplar trees can remove chlorinated solvents in groundwater. Alpine pennycress draws in zinc and cadmium. Sacred datura takes up lead. Cabbage can store radioactive materials–as can sunflowers.

In experiments in Chernobyl, densely planted sunflowers absorbed 95 percent of radioactive strontium in a pool near the leaky reactor. In 1996, the US secretary of defense and the Ukrainian defense minister sprinkled sunflower seeds over a former missile silo. Planting sunflowers also became popular in Japan in the area around the Fukushima nuclear power plant, where the 2011 earthquake left radioactive cesium and other toxins in the soil. Tens of thousands of sunflower seeds were given away by a grassroots citizen science group, and hundreds of thousands of sunflowers grew up in a newly created landscape of blazing bright-yellow flower faces. No one thinks now that these sunflowers had any real effect. The scope of radiation was too great and the sunflowers not planted densely enough. No one thought clearly enough, either, about how to get rid of the plants, which eventually became radioactive themselves. But other more systemic phytoremediation efforts are being considered.

My Nature's Notebook walk circles home again. Preliminary data from Nature's Notebook shows that deciduous trees around the country are leafing out early, with lilacs leafing out extremely early. More specifically, the overlap time is narrowing between the leafing of certain trees and the migrating Tennessee warbler. The Environmental Protection Agency and the National Climate Assessment committee will be using this information in their reports.

September 18

Two first instar holes have closed, those two larvae now quiescent in their burrows, molting into second instars. I dig out the remaining two first instars and preserve them in alcohol to send to Barry, having given up on the idea I will describe them myself. I have had less luck raising larvae this year, with fewer specimens, and I don't want to waste this opportunity to add to the scientific record. I've seen how much better Barry is at measuring things under the microscope, and the decision feels like a no-brainer, like using both sides of the middle-aged brain instead of one.

September 19

No larval burrows at the Gila River Bird Refuge, my regular spot, where I saw thousands of tiger beetles in July, and none at a nearby campsite, where Peter collected his beetles in the grass and sedge, seven in one swoop. I peer at flat areas of dirt along the water's edge (just in case these larvae do survive flooding) and then thirty, forty, fifty feet away. Soon I am out of the river's channel and into the cottonwoods and sycamores and ash, attracting stickers and seeds, the forest floor covered with leaf litter. I walk farther up, climbing now to the benches above the river. No tiny almost-perfect circular holes but–looking down and walking slow–I do find pot shards, bits of brown clay from a group archeologists call the Mogollon culture who lived here from 400 to 1100 AD (the start of the Dark Ages in Europe to the first Crusades).

Once I see shards, I want walls, a line of rock barely above the ground, or the depressions of a pit house fallen in and silted over. Some 1600 years ago, round pit houses here were dug three feet into the earth with walls of mud and poles and a roof of poles, mud, and grass. Women (most likely) made plain and polished red ware, waterproof clay pots superior to the previous woven baskets or gourds. This container protected food from pests and could be put directly on the fire, reconstituting dry food and simmering meat with greens and herbs and roots. Eventually these women started painting some of their pots, using red-on-brown and red-on-white designs. Later pit houses became stone-lined and rectangular, with villages of multiple pit houses that included kivas or ceremonial chambers. More settlements grew up, like the one I am exploring now.

As the labor of women increased–more food, more people, more crops to grow, more pots to make–cradle boards designed to carry a baby on a woman's back changed to cradle boards that could be set on a surface or hung from a beam. In rock and ceramic art, women and men are seen at different and specific tasks. In architecture, rectangular rooms allowed people to more easily set aside space for extended projects. Circular pit houses may have been preferred for their warmth and used mainly in winter by semi-nomadic groups. But by 1000 AD, the Mogollon culture had shifted to one-story, multi-roomed houses made of stone and built above ground, with large rectangular kivas nearby.

Also from this period, 1000 to 1150 AD, the interior of clay pots served as canvases for artists (also likely women) inspired by a new cultural renaissance. Narrative and fantastical images, along with geometric patterns, were painted in black on white: a creature half bighorn, half snake; a wolf wearing a deer mask; a woman giving birth; a man with a penis that had a little face–a little face sticking out its tongue! Today these pots sell for hundreds of thousands of dollars, and for this reason, without doubt, this small site has already been illegally pot hunted, mostly by shovel but probably with larger equipment since the area is so accessible to a road. Pot hunters look for the pots and funereal goods often buried under the floors of homes, buried with the dead, whom many Native American groups see as journeying until the last bit of bone and pot crumbles. To disturb these graves is to disturb these journeys. Moreover, to disturb archaeological sites on public land is illegal, punishable now by fines and imprisonment.

I happen to know that archaeologists have installed motion-sensor devices at certain sites at the Gila River Bird Refuge, which I might well be walking through now, my movements triggering a text message sent to a satellite phone monitored by an archaeologist who will stop by later to see if I have caused any damage. It is not illegal or wrong, of course, for me to be on this river bench on public land–as long as I am careful not to dig up the ground and to put back, as well, any artifact I find.

I pick up shards, rub their edges, feel their textures. I admire a bit of burnished red, sensuously curved, the lip of a pot. I like as much the inch-long square of a corrugated cooking vessel, bumpy and rough. The thrill never gets old for me, this making of pots, this life by the river. I feel the sun on my neck, the smell of dust, my heart beating its pulse of blood. Like her, like this potter, I am another animal in the landscape. Maybe I have children. Certainly I have worries. There is danger, from injury and accident, from lions and snakes and other humans. I might get an abscess in my tooth. I might have a difficult mother-in-law. I'm working hard to survive, but that feels normal, while my other life in the twenty-first century is suddenly dreamlike–an amazing, amusing, fantastical dream. I have sometimes stopped in a store, in a restaurant, at a traffic light, confounded by an atavistic awe, both appalled and appreciative. Wow, I think. Who did that? What happened to the trees? I have felt, as I do now, alien to the modern world. A time traveler not born here, but born there, not at home here, but belonging there.

We're so flexible, we humans, with our plastic minds. Almost anything can begin to feel normal, from living in a spaceship to looking for larval burrow holes, those tiny almost-perfect circles in the ground. I put the shards back exactly where I found them and mentally text any archaeologist in range: 2BZ4UQT.

September 23

When you're a hammer, everything looks like a nail, and when you're looking for the larval burrow hole of a Western red-bellied tiger beetle, you see a surprising number of holes you've never seen before. Usually they are not the right size or shape, but you look inside anyway because you want to know: who lives in all these holes?

A number of circles, eighty feet from the riverbank where I have seen hundreds of adult tiger beetles, are too large but still irresistible. Hole after hole, nothing lives there now. Instead something has emerged over the summer.

Tiny perfect holes in the dry upland grass seem more promising. Tiny ants are passing by, and for the first time I can see how tiger beetle larvae might survive in the wild—as I imagine these tiny ants coming too close, and the tiger beetle larva lunging out and grabbing its prey. Then I notice how often the ants are marching into these holes, which are obviously their nests.

Other holes near the trail have turrets or small mud chimneys. I don't bother to look inside these, knowing they were not built by the Western red-bellied tiger beetle. Possibly they are the old nests of digger bees whose turrets prevent parasitic flies from flipping their eggs into the burrow to hatch and devour the bee larvae. Similarly, Williston tiger beetles construct their turrets on Laguna del Perro and other salt flats.

Closer to the water, in dry cliffs that once marked the river's channel, I see lots of cicada emergence holes and what I think is the home of a tarantula. Tarantulas start their burrows as spiderlings and live there a lifetime, as long as ten years if male and twenty-five if female. This entrance is over an inch in diameter and covered with a light veil of silk that keeps in humidity and carries vibrations down into the foot-long tunnel with its J-shaped chamber. About three inches long, tarantulas hunt beetles and grasshoppers and other small prey at night. Their defense against the foxes and coyotes and raccoons who like to eat them are irritating abdominal hairs that fall off easily and get into a predator's eyes or nasal passages. (Coatis have learned to dislodge those hairs by vigorously rolling the spider back and forth along the ground.) Most people who walk around the Southwest become fond of tarantulas and think of them as lucky, much like having a roadrunner cross your path. In this case you have all the time you need to stop, point to the large, hairy, slowly moving tarantula, and call out to friends who can gather around and watch with you.

Along the river now, there are signs of beaver chewing on tree trunks and sliding into the river; perhaps there's a den nearby. Southwestern beavers tend to make bank dens rather than lodges, a bank den having several entry tunnels with one above the high water mark. Its single inside chamber is about two by three by three feet. Other holes I'll see on this walk might be made by gophers or ground squirrels, pocket mice or grasshopper mice. Collared lizards and whiptails use the holes made by other animals but occasionally dig their own burrows with a half-inch, half-moon shaped entrance. Wintering snakes also borrow someone else's hole and sometimes den communally, rattlesnakes and bull snakes and whipsnakes together. Burrowing owls modify the holes they find by lining the interior with feathers, food debris, and horse and cow dung. *A Field Guide to Desert Holes* says blandly, "This may be to disguise their scent to predators or as decoration." Similarly, skunks borrow burrows or make their own, decorating them with a strong musky odor. Coyotes only use dens when birthing and raising pups, often on a hillside or bank, the hole taller than wide. There are a few large mysterious holes on my Nature's Notebook walk that I like to think were made by a badger, a prodigious and powerful digger.

I guess we just see the top half of life.

September 29

A friend brings by a mating pair of Western red-bellied tiger beetles he found in the Burro Mountains near a stock tank, not far from where the Sky Island Alliance team looks for tracks. He says the beetles were fairly abundant, two or three every square feet. I saw a few tiger beetles along the Gila River through the end of August, and David Pearson reports them in Arizona in September. Still, this seems late in the season for this area.

Peter and I drive out to find the stock tank but get lost instead. The national forest here is crisscrossed with old mining and logging roads, and although we have a four-wheel drive car, we still slip and slide when the road turns to sand, and so this becomes a long day, nerve-wracking and familiarly Western, lost and skidding, surrounded by undulating juniper and scrub oak–overgrazed land, dry land, scarred land.

Another road becomes an arroyo. We get out and walk to the end, a boxy canyon rising up in pink and white slick-rock. This isn't the way to the stock tank either, but we do see lots of bear track. The prints go up and down, up and down, passing each other like hobbits. Another bear highway. By now I've seen bear tracks on a number of walks I do all the time along the Gila River, near Silver City, and once at the Nature's Notebook trail in my backyard. My epiphany with bears is that, huh, there's a lot of them.

At last Peter and I find our way home, and I am left thinking about the seasonality of tiger beetles, still congregating around water holes in the Burro Mountains. Can these be part of the same mass emergence of adults who appeared along the Gila River in early June–at a similar elevation, twenty miles to the west? The tiger beetles I captured in July and August died in late September, and I had assumed their lives had been unnaturally prolonged in my terrariums. But maybe not. Or maybe batches of adults emerge at different locations at different times throughout the summer.

First, the larvae have to overwinter successfully, quiescent, waiting. They have to time themselves to pupate together so as to emerge together and find a mate. Conditions for fertilization have to be right: enough water, enough food. Adults may or may not need to disperse, leaving flooded or dry habitat, finding someplace new. They have to adjust. They have to adapt. They have to persist. This species is so flexible.

Pot shards (photo by Marilyn Gendron)

October 2012

October 3

No larval burrow holes.

October 6

No larval burrow holes.

October 10

No larval burrow holes.

October 13

I e-mail David Pearson that I don't think I'll see open larval burrow holes until after a rain, which would mean food: a greening and flurry of prey. Certainly the larvae in my terrariums open and close their tunnels in response to water.

In those terrariums now, I have two second instar larval holes and three more first instar larval holes. I would like to rear up another third instar for Barry to describe, as well as send him some second instars.

October 20

My friend Carol and I bring the Mastodon Matrix Project to the Gila Valley on a Saturday, when kids from the local school as well as kids who are home-schooled can attend. We set up in the Gila Valley Community Center, a somewhat grim bare-bones room. About eleven children arrive, with five parents, and Carol talks about mastodons dying and sinking to pond bottoms in New York. Then we plop down scoops of fossil matrix on paper plates. Soon exclamations and cries fill the air. "I've found a shell!" "Is this a tusk?" "Where do I put rocks?" "Is a leaf a plant?" One fourth grader is particularly intent and keeps sifting through his chalky gray matrix long after the other children have gotten bored and gone outside to play. He is working even as we clean around him, sifting and murmuring to his mother who looks on patiently. "Yes, he's always like this," she tells me. Before they leave, the boy asks Carol, "Are we doing this again next Saturday?"

A new and similar citizen science project is Fossil Finders, co-sponsored by the Museum of the Earth in Ithaca, New York and the Cornell University Department of Education. The scientific question is "How did the organisms in the shallow Devonian sea of upstate New York–416 to 359 million years ago–change in response to environmental changes?" Guided by a teacher, middle- and high-school students are sent rock samples rich with fossils, mostly shelled animals. Again this original research involves authentic materials. As the website explains, "You and your students will be the FIRST PEOPLE ever to see these fossils!" A dichotomous key helps the students identify what they find.

> 1. My fossil has a whorled body or shell. If Yes, go to 2. If no, go to 1b.
> 2. 1b. My fossil has an elongated body or shell. If yes, go to 3. If no, go to 1c.
> 3. My fossil is very straight, gets narrow at one end, and larger at one end, and has lines or chambers. If yes, go to 4. If no, go to 5.
> 4. If your fossil is straight, narrow at one end and wider at the other, and has chambers, you have a straight cephalopod!

The information is entered into an online database. Although the point is for students to learn science, as with the Mastodon Matrix Project, some real science will get done along the way.

Today I also log on to the National Geographic Valley of the Khans Project, which has been using crowdsourcing to uncover the tomb of Genghis Khan. National Geographic calls this brutal thirteenth century ruler "the most accomplished man to have walked the earth" by virtue of having conquered most of central Asia and China. Khan additionally introduced an alphabet and currency into his kingdom of warring tribes, and one in every two hundred men is said to be related to him.

The most accomplished man to have walked Earth really, really, really didn't want his grave found. Legendarily, his funeral escort killed anyone they happened to meet, killed the slaves who buried Khan, killed the soldiers who killed the slaves, trampled the ground with ten thousand horsemen, planted a forest over the site, and then diverted a river to cover their tracks. Some of this may have happened later as his successors chose to be entombed with the conqueror, adding to the funereal treasure.

At the California Institute for Telecommunications and Information Technology, Albert Lin and his team have been looking at eighty-five thousand satellite images of the mountain range where Khan's remains are likely to be found. They are helped by thousands of online volunteers, who identify modern structures like roads and corrals, natural features like rivers, and interesting odd formations that could be archaeological sites. Enough interesting "tags" by these citizen scientists and Lin's teammates get on their horses and go exploring–or "ground-truthing." Three promising sites have already been tagged. One was from the Bronze Age, more than three thousand years old–too old. One was also an earlier tomb that had already been looted. The third seems to be the site of a thirteenth century temple with signs of a large man-made object under the soil. Before excavation, the team now needs permission from the Mongolian government.

On the website, I drag icons of rivers, roads, modern structures, and ancient structures to a satellite image, advancing in half an hour from Novice status 1 to Novice status 2. I am often encouraged: "Well done, Sharman! You've tagged two out of five important features." After a while, I get into the groove. Whenever I feel overconfident, they throw me something difficult–a fuzzy image, a new pattern. When I fail, they encourage rather than scold: "Nothing to see here? Sixteen previous explorers saw a road on this map. Do you agree?" This is much more fun than Angry Birds.

Another historical project, Ancient Lives, wants me to transcribe some five hundred thousand fragments of Egyptian papyri uncovered in 1896 in Oxrhynchus, or the City of Sharp-nosed Fish. Ancient Lives is part of Zooniverse, which began with Galaxy Zoo in 2007 but has since proliferated into a dozen programs that rely on human numbers and intuition. Besides cataloguing the shape of galaxies, citizen scientists in Zooniverse can explore the surface of the moon (they've already analyzed over two million images from NASA's Lunar Reconnaissance Orbiter), look at the weather on Mars (studying wind patterns on the Martian surface using data from the Mars Reconnaissance Orbiter), find planets around stars (mapping light curve changes and information sent back by the Kepler Spacecraft), and determine how stars form (examining infrared images from the Spitzer Space Telescope). They can help model climate change by deciphering the handwriting of World War I-era navy captains–looking at weather observations from scanned ship logs–or by analyzing thirty years of tropical cyclone data. They can classify bat calls or cancer cells. They can identify ocean species as they watch images of the seafloor along America's northeast continental shelf–and right here, I stop and log on again.

The philosophy of Zooniverse is that people are smart. And impatient. They don't need a lot of introduction. Immediately I see a digital photo taken by a camera system clicking six images per second as it is dragged three to nine feet above the ocean bottom. In this way, over thirty million images have been collected in less than a year. My training photo is the top view of a fish on a sandy bottom, with some scattering of white shell and a few fan-like shapes. First I identify the bottom surface: sand, shell, cobble, boulder, or can't tell. Then I am given four categories of living organisms–fish, scallop, crustacean, and sea star–and shown how to use my cursor to measure the dimensions of what I see. The fan-like shapes are scallops, and I am told not to record the dead one with a hole in the center.

Before leaving the photo, I'm asked if I see other species. I can click yes or no, or go to Help for any niggling problem, which is where I learn about common animals I might encounter, like the invasive sea squirt and sea pen. Also, I'm told to count all brittle stars, basket stars, cushion stars, and sun stars as sea stars. What's the difference? I pause the oceanography to go to Wikipedia. Then I'm on my own, image after image, sea star after sea star, a scallop, another scallop, rarely a fish. Sometimes the ocean floor is just sand. Sometimes I see a community of life in close detail, three feet away–two scallops, three sea stars, one sea urchin. I never know what I will see next. I click. I go to the next photo.

In less than an hour, according to Sea Floor Explorer, I have looked at twenty-three images and counted forty-one sea stars, twenty-one scallops, one fish, and zero crustaceans. My colleagues in this project have looked at over one million images and totaled almost two million sea stars, one million scallops, over eighty-four thousand fish, and over seventy-eight thousand crustaceans. The questions scientists want to answer with this data are: "What is the current distribution and abundance of sea scallop and yellowtail flounder on Georges Bank, and how have they changed over the years?" and "Where is the invasive sea squirt currently co-located with gravel substratum, and what is the potential for its spread to new areas?" Suddenly these are exciting questions and of concern to me, too.

Before leaving Zooniverse, I also visit Galaxy Zoo, something I've avoided in part because my expectations are so high. We live in a spiral-shaped galaxy with an estimated two to four hundred billion stars in a universe with as many as five hundred billion galaxies, a number that makes me want to take a nap. Galaxy Zoo is looking at galaxies–images in a database–that no human has ever seen before, images that take billions of years to reach us since their light is traveling across unimaginable space. Too much unimaginable space and time and I start hoping for a mystical experience; I'm not going to understand this otherwise.

After their first project looking at a million unclassified images of galaxies, Galaxy Zoo asked volunteers for more details on two hundred thousand of the brightest galaxies; how many spiral arms did a spiral arm galaxy have, for example, and how big was that center bulge? Their third project used images from the orbiting Hubble Space Telescope, and the current study is a refinement of that, looking at views from Hubble's latest infrared camera, installed in the final shuttle mission.

Galaxy Zoo does not mince words. Log on and prepare yourself: "Few have witnessed what you are about to see." Moreover, "If you are quick, you may even be the first person to see the galaxy you are classifying." Next is my first digital photo. Is the galaxy smooth and rounded with no sign of a disk? Click one of three possibilities, each with a picture. Smooth. Features or disk. Star or artifact. (What's an artifact? I wonder. Go to Help for this.) Could this be a disk viewed edge-on? (What does that mean? Go to Help for this.) If so, does the galaxy have a bulge at the center? If yes, what shape? Click the right image. Is there anything odd? (It's all odd, I think.) Would you like to discuss this picture? (Go to a discussion forum.)

For the moment, the most famous citizen scientist at Zooniverse is Hanny van Arkel, a Dutch schoolteacher who did see something odd next to a spiral galaxy and who did go to a discussion forum. Other people had noticed this glowing cluster, too, but hadn't pursued it. Hanny kept asking. The object is now known as Hanny's Voorwerp, a gas cloud the size of the Milky Way illuminated by a quasar or light beacon from a black hole.

Hanny is on record saying "it's pretty cool" to have a vast area of space 650 million light years away named after you.

You look for the next Voorwerp and hope for galactic fame. But mainly you are helping build the huge samples of classified galaxies needed for research, whether it's a question about elliptical galaxies far away or spiral galaxies closer to home. During my online session, I look at one photo from the orbiting Hubble Space Telescope and eleven from a telescope in New Mexico. My classifications are redundant, multiply checked against those of other humans looking at the same digital image. I'm certainly glad for that.

Finally, I take a survey—with a chance to win a gift certificate at Amazon.com—that tests my knowledge of galaxies and space. The researchers at Galaxy Zoo want to know what their volunteers know. In my case, we are both appalled. The questions assume a surprising familiarity with physics and include diagrams. Clearly, the larger community at Galaxy Zoo is well informed. Their sophisticated human eyes are scanning images across unimaginable space and time, watching galaxies interact, contemplating dark matter, dark energy, and the 5 percent of the universe that remains. Their computer lives are rich and full, with plenty of razzle-dazzle if not a mystical experience—and likely some of those, as well.

And I'm sold. I'm addicted. I've got to add one more citizen science project to my life. But it won't be an online Zooniverse project. As a writer and teacher, I already spend too much time at my desk. Joining the Sky Island Alliance tracking team in the Burro Mountains is tempting, but for now a Saturday every six weeks is too big a commitment. Instead, I'm looking into New Mexico's Site Steward program, in which volunteers go to archaeological sites in the Gila National Forest and monitor them for damage. After the initial training, I'd visit my site every four to six months.

I call my connection—ah, an archaeologist in the national forest—and we make a date to meet. Razzle-dazzle.

October 24

No larval burrow holes. After tramping the banks along the Gila River Bird Refuge, I extend my search farther into the cliffs above the river, not expecting to find anything here since the ground is so rocky and dry. This year is shaping up to be New Mexico's second driest year on record. By mid-October, we've had less than six inches of rain in the Gila Valley, with slightly more in Silver City, and slightly less elsewhere. Perhaps I should stop using my Saturday and Wednesday afternoons to look for Western red-bellied tiger beetle larvae and just wait until it finally does rain, if it ever does rain, if it ever rains again–except that tiger-beetle-larvaling remains such a good excuse to get out into the world.

I don't always go alone. I take my friend Madge. I take my friend Shirley. Maybe we talk more than we actually look for larval burrow holes, talking and talking about our lives and our children and what we think our elected leaders should do to fix the world, a world we've fixed so many times in these conversations that I'm always surprised when I hear the news again the next day. When I do venture out on my own, I enjoy that too, and I know this happens because I am here with a purpose. I'm not just a lonely person without friends. I'm looking for something. I have a task in the natural world, climbing up and down dry dirt banks, searching the ground.

Today, walking back to my car–but still with purpose, glancing about–I see a pattern of scratches four feet up the white, papery, peeling trunk of a sycamore. The five parallel lines are an inch apart, too big for a bobcat. I don't see any tracks below, and the scratches don't seem fresh, although I do feel suddenly alert. I do turn in a slow circle.

Mountain lions scratch trees much like house cats scratch furniture, standing on hind legs and dragging their claws down. This may be part of lion grooming, keeping those weapons clean and sharp. The paws also leave behind scent, and the scratches probably serve as an advertisement for sex or territorial marking. The home of a mountain lion is large, depending on the availability of food and whether the lion is a male or a female with young or a female without young–and although males and females can share some overlap, males cannot and must be clear about where the lines are and who is crossing them.

In the wild, mountain lions can live ten years or more. That's a long time to get to know a place, a long time to leave your scratches as well as your many piles of scat and urine, the various signs of yourself. In your rich and full life here on the Gila River, you've found and pursued mates with an intensity that defines you. You've eaten hundreds of deer. Maybe you've raised a number of litters. Maybe you killed another lion or saw your cubs killed by one. Mostly you've lounged, and often you've drowsed. You've been here forever, dreaming, scheming. You've thought about—well, you don't really have thoughts, not that we know about, the little we know about you, the mystery that you are.

I lean back and stare up at the tree. As a writing teacher, I tell my students: don't, please, write that something is "indescribable" or

Scratches on sycamore (photo by Elroy Limmer)

"inexpressible." To describe, to express, that's your job. But the sensuous tangled limbs of this sycamore–so much like human arms rising up in dance–the whiteness of the bark saturated and rich, giving off a kind of light, a luminosity, the entire tree so clearly a goddess ... all this seems indescribable.

I feel a movement in my ribcage. My chest hollows.

And this, I see, moving on and looking at the ground for larval burrow holes, is how I make my own mark, turning the world into beauty with my comment, my claw, the various signs of myself.

October 25

A boy bangs his fists on the table, over and over. He can't seem to help himself. Apparently his mother didn't give him his medication this morning because she was fighting with her boyfriend. Another child in my daughter's third-grade classroom slides from his chair and drops to the floor. He is an ESL student who struggles with his work and seems constantly angry. Across the room, two other boys are loudly conspiring. A little girl sings to herself. She's in a bubble. Sixteen children are being quiet, but seven are not, so that the room feels chaotic and unhappy. My daughter is trying mightily to get these eight- and nine-year-olds to listen to her instructions about our project Urban Birds from the Cornell Lab of Ornithology. But this is a far more difficult class to manage than last year's, with too many troubled students and too many behavioral problems dumped on teachers like Maria. I'm exhausted by the effort she has to make every minute, every hour, every day, just to get everyone's attention. If only she had an aide. If only she had parent volunteers. If only she had eight eyes and twelve hands and could bilocate. Finally, we go out to the playground for ten minutes to look for birds. Ten minutes have never seemed so long. Twenty-three children have never seemed like so many children. And we do not see a single bird. Not one. Now I understand the virtue in Cornell Lab of Ornithology's insistence that Zero Means a Lot! Data is data. Not seeing anything is still data. We hand out the stickers Zero Means a Lot! We tell the kids not to feel badly that there weren't any robins or pigeons or mourning doves or grackles today. (Do feel badly, I think as I drive away, almost weeping, at how the educational system is failing these children. They and their teacher deserve better.)

October 29

For my first Site Steward training, Bob and Elizabeth–two archaeologists with the Forest Service–meet me at a designated spot, and then we take their impressive government truck. On the dirt road up a steep hill, we scare a bighorn sheep, its curved horns bouncing away into rock and juniper. Bighorn sheep are fairly common on painted bowls from the late period of the Mogollon culture, which may mean they were a common food source or important culturally or just visually interesting. The growth of a wild sheep's horn follows a mathematical series in which each number is the sum of the two previous numbers–similar to other spirals in nature, nautilus shells and sunflower seeds and galaxies. After the fourteenth number, every number divided by the next highest results in the length-to-width ratio of what we call the golden mean, the basis for the Egyptian pyramids and Greek Parthenon and for much of our art and music. In our own spiral-shaped cochlea, musical notes vibrate at a similar ratio.

In the Southwest, the population of bighorn sheep neared extinction by 1940 but rebounded with a reintroduction program in which the animals are reared in isolated areas and transplanted. Native desert bighorn are released in southern New Mexico and Arizona, but the Gila Valley was initially stocked and remains so with a related species, the Rocky Mountain bighorn sheep. Possibly this very sheep came here in a truck, although more likely his parents or grandparents did.

Later in the day, walking through mesquite and prickly pear, I start seeing signs of looting, dozens of recorded excavations or potholes. Most of the illegal excavations are centered around the remains of collapsed stone walls, part of above-surface room blocks from 1000 to 1150 AD. Before this village, earlier groups settled here, perhaps in pit house villages just as large. The ground is covered with their pot shards, too: red on brown, corrugated plain ware, red slip, brown ware, a thousand jigsaw pieces, and none of them match. Innumerable pots, lives, dramas. Today Elizabeth will extend the range of occupancy, previously thought to be from 800 to 1200 AD, when she sees polychrome shards from the Salado culture–a group who lived in the upper Gila from about 1375 to 1450 AD, their villages characterized by large pueblos often made of puddle or coursed adobe. The Salados usually cremated their dead instead of burying them,

Bowl with two bighorn sheep, human head and figure; Mimbres people, Mogollon culture, New Mexico, c. AD 1000–1150 (Dallas Museum of Art)

and their ceremonial rooms don't look like recognizable kivas. Probably these people were migrants from the west or north. Possibly they evolved from local groups. Conceivably, this village was a cosmopolitan mix, multilingual and multiethnic.

I pick up (and put down) a square of white with thin black lines, part of a bighorn sheep or crane or mountain lion. I pick up (and put down) a curved piece of corrugated brown the size of my palm—all those stews, simmering meat and roots and herbs. I pick up (and put down) a geometric pattern of red and white, the imposition of rectangles and squares and parallelograms. A human shout: look at the world in *this* way. I pick up, I pick up, like a child on an Easter egg hunt. My metaphoric basket. All the treasure I have metaphorically found.

Bob pauses at an excavated hole that looks fresh. "See how the lip of the hole isn't rounded much or worn down," he says. "And there's not much debris inside that would normally be carried there by wind or animals." He think it's worth having one of their special agents inspect the damage, perhaps find some clues. "Careful where you walk," he tells me. "It's a crime scene now."

This is the job of a site steward, explained in the Site Steward Handbook: Find It, Record It, Report It. The program is set up to avoid confrontation between its volunteers and any potential looter or criminal. Approaching our designated site, we should move quietly and cautiously, muting any cell phone or radio. If we see anyone on the site, we should watch them from a distance, "safely and quickly collect whatever information is possible," such as license number and "subject description," and then we should leave. The manual warns, "Never place yourself or your vehicle on a hilltop or on the skyline of an open ridge–this makes you easy to spot. If there is no obvious way to conduct your observations safely–do not observe!" We are also reminded that the sun reflecting off binoculars can signal our presence and that we should dress for weather, carry plenty of water, tell our site manger when we are going, travel in teams if possible, watch for natural hazards, not pick up litter (it could be one of those clues), gas up our vehicle before leaving, stay near the vehicle if it breaks down, be careful while driving through arroyos, and avoid prolonged exposure to sunlight. Sending people out to remote areas in the Southwest is no joke. But the fact that this site has been nominated and approved for volunteer monitoring means it has some importance. This once large and long occupied village, where we are standing now, has subsurface structures that new technology could explore, with more to say about these cultures, their trade networks and settlement patterns.

The location of this site is a secret, of course, although I can think of a dozen people I'd like to bring here. Instead I'll have to come alone or with another steward, approaching these swells of juniper and mesquite quietly and cautiously, filled–and this is my personal instruction, added to those from the handbook–with the kind of ego displacement that comes from thinking about history and the sweeping pattern of lives come and gone.

Bob, Elizabeth, and I visit three other sites I will also monitor: a small stone-masonry surface structure; a large room-block, heavily pot-hunted, and a single-room cliff dwelling–high and hidden on a crumbly slope that requires some climbing. The narrow entrance to the cliff dwelling still has its wooden lintel. The adobe wall blocking this ledge is still solid. Peering into that darkened room, hand on the lintel, I feel the frisson of time-travel scented with mice urine. Elsewhere on the ledge, we find quartz crystals wedged into porous rock, and Bob speculates that this might have been a shaman's camp, used ceremonially. But he doesn't know when. People were doing a lot of that in the 1960s, too.

As we walk back down a slick-rock canyon, I find myself ahead of the other two, the streambed only a few feet wide, pinkish white-gray rock above my shoulders. Now I am passing a cliff of yellow and orange, weathered and oxidized, showing a bit of iron before the canyon narrows again. I'm about to turn a corner, touching the granite wall with one hand. The flow of water, its rushing presence, is still here in the way these boulders pile against each other, creating this narrow space, flood debris caught in the cracks and angles, tangles of root and dirt and stone. I'm about to turn a corner like any corner, the air rich with oak leaf and the musk of some animal. The world is pink and white under the sky's blue bowl. I break the silence with my own steps. What track, what sign, what gift, what next?

Sandhill cranes (photo by Elroy Limmer)

November 2012

I empty the terrarium of instars and put what I have–four first instars and one second instar–into vials of alcohol to send to Barry to describe.

Ten years ago, another entomologist, Dick Vane-Wright, Keeper of Entomology at the London Museum of Natural History, inspired me. *You could spend a week studying some obscure insect and you would then know more than anyone else on the planet.* There's still so much to discover. We think we've beaten the Earth flat, hammered out the creases, starched the collar, hung her up to dry. We've turned the planet into our private estate, no longer wild, no longer mysterious. And yet. *You could spend a week studying some obscure insect and you would then know more than anyone else on the planet.*

And maybe that "you" is now me–if we count intangibles. If we count admiration, watching a pack of Western red-bellied tiger beetles as they scythe their mouthparts and dice up prey and leap on mates, as they stilt and stare, as they run, pause, look. Run, pause, look. That fierce commitment! Grabbing hold. If we count worrying, finding just the right soil for the shape, thin or broad, of the female ovipositor; adjusting heat lamps; re-ordering mini-mealworms. If we count stalking and killing, dismembering bodies. If we count ovaries. If we count familiarity. A comfortable recognition–those seven creamy irregular dots. If we count surprises. Why this sudden temper? Why here on my gravel path? Why close your burrow hole so soon? And why–why, look at you, you splendid little ugly creature marching up and down on my kitchen table.

Maybe that "you" is now me if we extend this relationship to the rest of the world, the back roads of New Mexico, the steps of the National Museum of Natural History, seven million beetles in metal drawers, larval hoops, the pygmy skittering on Dick's lawn, the way the blue jay cocks its head, the way the Gila River floods and ebbs and floods again, the red coin of the sun, smoke in the air, scratches on a tree, the physics of beauty.

Maybe that "you" is now me if you think "Barry and me." Because once Barry gets the larvae I reared and puts them under a microscope, we'll know at last how many setae are on the antennal segments of a first instar's head. We'll know the number of median hooks on the fifth abdominal segment of the second instar. We'll have a description of all three larval stages of the Western red-bellied tiger beetle.

And maybe that "you" is now me since when I do find larval burrows in the wild–this year or next year or the next–I can compare those larvae to what was raised in the terrariums, as can any interested researcher, and we will replace that phrase "larval biology unknown" with some specific information as to whether these females prefer the sides of grassy hummocks or moist clay-like soil or shallow drainage swales. We will fill in that blank spot on the map of tiger beetles.

Of this, I have no doubt. I know when to look, and where to look, and for what. Once I find those sites of nearly perfect circular holes, I'll be able to do a little excavation, see which stage of instars are overwintering, see when the adults emerge. If I choose to raise more larvae, I know certain tricks, and Barry and I can increase the number of instars for a better description. Maybe I'll learn to describe them myself.

I'm deeper into the world of the Western red-bellied tiger beetle and I'm deeper into my own, living here in southwestern New Mexico, taken deeper by citizen science programs like Nature's Notebook and Sky Island Alliance and Site Steward and Galaxy Zoo–even Galaxy Zoo, especially Galaxy Zoo, looking up at the Milky Way, one of the arms of the spiral galaxy where I live, one of over a few hundred billion galaxies, some of which I may have been the first person to ever see.

Where is my window into the unknown, the nonhuman? And where is my competence? My expertise? My forest. (My Gombe.)

Oh, right here.

You could spend a week studying some obscure insect and you would then know more than anyone else on the planet.

* * *

At the end of November, the end of my second season chasing tiger beetles, my friend Tris takes me up in his small plane, a Cessna 47. We go north from Silver City, meeting up with the course of the Gila River, turning at the edges of the Whitewater-Baldy Fire, heading back over my house and my neighbors' houses in the Gila Valley. My ninety-three-year-old father-in-law sits in the back seat, quite enthused. A former officer in the US Army, he has been in plenty of planes and even jumped from some of them, but that was a long time ago. That was then. This is a new adventure.

Minutes into the flight, three million acres of mountains and ponderosa pine and juniper rolling beneath us, and I am in love with flying. "I should have done this," I tell Tris, meaning all he has done—getting the pilot's license and buying the small plane and putting in all those hours of preparation and work and adrenaline.

"It's not too late," Tris points out.

"Why not?" my ninety-three-year-old father-in-law agrees.

I file that away.

Clouds mass and billow. The landscape unfolds, undulates, brown and green and a line of gold along the river, the yellow of cottonwoods, the russet-red of sycamore. We fly over the Gila River Bird Refuge where next summer the monarchs and netwings will hover over the white-flowering clover bush and the foxes and coatis and black bear and lions will be signaling to their young—sweethearts, follow me—and thousands, thousands upon thousands of Western red-bellied tiger beetles will be clamoring the mud and sandy banks, in the sedges and grass, along the edges of water. Skittering, darting, flying, pouncing.

I can see them still, an abundance of life.

Sandhill crane (photo by Elroy Limmer)

Epilogue

Thirty years ago, my midwife was seven months pregnant when I was nine months pregnant, and today at a party I am talking to her son, who is getting his PhD in molecular biology at New Mexico State University. He is working on a microbial vaccine for chili plants, trying to induce an immune reaction for a better defense against blight disease. He asks what I am writing about these days, and I say citizen science, and he asks what I have learned, and I say, "Well, I now know why I didn't become a scientist," and he bursts out laughing.

"Yes!" he says, "Everyone thinks it's so glamorous and romantic." I am laughing, too, "But there's so much detail work!" I complain. "You have to quantify everything. You have to measure everything!" We bad-mouth the tedious aspects of science, gathering data, inputting data, and the caution required–trying to say something simple about a complex world–but it's family talking about family: you can make fun of your grandmother because she's your grandmother.

I tell my midwife's son about tiger beetles, the little I've learned, and he tells me about the microbe *Phytophthora capsici* and related species, the little he's learned. (There's so much to discover.) He's heard about some of my favorite citizen science projects, and he personally logs on to InnoCentive, the online site where "creative minds solve some of the world's most important problems for cash awards up to $1 million." Those problems range from new ways to get energy from algae to better material for surgical gloves. I'm intrigued anew by this mixing up of science and citizen science and the entrepreneurial spirit. I'm reminded again: the power of citizen science is not going to be kept in a tidy box. The potential of citizen science will still surprise us.

Today, the typical citizen scientist in America is white, well-educated, and middle-class. More outreach needs to be done. Even so, the field of citizen science is inherently democratic, offering opportunities for almost everyone in almost every scientific discipline. You can be an auto mechanic designing medical equipment or a third grader in Deming, New Mexico filling out her observation of a robin. You can be the first to transcribe a papyrus from the City of Sharp-nosed Fish or find a new species of fly in your backyard. You can transform yourself in a variety of ways. Become an expert in bryophytes. Experience Zen-like moments in the office and in the field. You can do public good–add to scientific knowledge, monitor changes in the environment, promote better social policy–even as you roam your private paradise, whatever and wherever that is, collecting treasure and bringing it home: crumbling seed pods, feathers in your hair, clouds in your pocket.

Along the way, send me an e-mail. Tell me what you find.

* * *

From: "C. (Barry) Knisley"
Sent: Tues, Jan 7 2014 9:59:38 AM
Subject: RE: Western red
Sharman, I have been thinking recently about various applications of these relatively new and inexpensive game cameras for tiger beetle studies. A colleague in TX is using one to observe activity and movement of Dromochorus. They can be put in place easily and with an available lens (<$200 I think) can cover a square meter or more of ground. I have not explored all possibilities of how useful they may be. But one of these might show us if the beetles leave the water edge at night. And of course, they would be invaluable to observe various behaviors for beetles in a limited patch of ground. Barry

Appendix
Description of the First and Second Instar Larvae of *Cicindela sedecimpunctata* Klug

Larval stages of the tiger beetle larvae of the Sulphur Springs Valley of Arizona were described by Knisley and Pearson (1984). All three instars of the fourteen previously undescribed species were included, plus the third instars only for *C. sedecimpunctata* and *Amblychila baroni*. Most of those described were from larvae that were reared from field-collected adults. Unlike most other species, *Cicindela sedecimpunctata* produced very few eggs and only one third instar could be obtained for description. Fortunately, as part of a citizen project with tiger beetles, Sharman Russell was successful in rearing larvae of this species from field-collected adults and obtained one second instar and four first instars that are used in the description presented here.

C. sedecimpunctata Larval Descriptions
Second Instar (Figure 1)
Measurements (in mm): (1 specimen) TL 8.37; W3 1.06; PNW
1.47 ; PNL ,83 FW 1.39 ; FL .80
Color: Head and labrum black brown; pronotal disc dark brown, cephalolateral angles. Antennae medium brown. Maxillae yellow to yellow brown; labium brown. Mandibles yellow brown, apices dark brown. Mesonotum and metanotum medium brown anteriorly, cream posteriorly. Dorsal cephalic and pronotal setae white, other body setae yellow.
Head: Dorsal setae prominent; diameter of stemmata 1 and 2 and the distance between them about equal; U-shaped ridge on frons with 2 setae, small conical projection between setae. Antennal segment 1 with 4 setae, segment 2 with 5-6 setae
Pronotum: Pronotal setae prominent, 6 pairs, s1, S3, P2, P6, and P7 absent; 23-24 pairs of lateral marginal setae, 13-14 pairs of cephalomarginal setae.

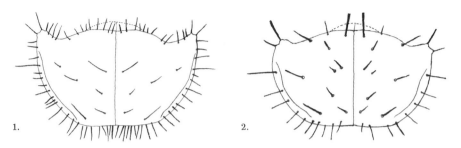

C. sedecimpunctata: 1. Second instar pronotum, dorsal aspect. 2. First instar pronotum, dorsal aspect.

Abdomen: Sclerotized areas distinct. Third tergites with 12-14 setae. Median hooks with 2 setae. Inner hooks with 2 apical setae, spine one-quarter total hook length. Fifth caudal tergites with (24-28) setae; epipleura with (3-5) setae. Ninth eusternum with 2 groups of 3 setae on caudal margin; pygopod with 6-7 setae on each side.

First Instar (Figure 2)
Measurements (in mm): (4 specimens) TL 6.51; W3 .74; PNW 2.27; PNL .61; FW 0.99; FL 0.61
Color: Head and labrum black; pronotal disc and cephalolateral angles dark brown; antennae yellow to brown. Maxillae and labium yellow brown, apices black. Mandibles yellow brown basally, apices black. Mesonotum and metanotum light brown anteriorly, cream posteriorly. Dorsal cephalic and pronotal setae white, other body setae yellow.
Head: Dorsal setae prominent, numerous pairs (bifurcate at tip); U-shaped ridge on frons with 2 setae, small conical projection between setae. Antennal segment 2 with 2 setae.
Pronotum: Pronotal setae prominent, 7 pairs, s1 absent; p6 short, p8 slightly bifurcate at tip; 9-11 pairs of lateral marginal setae. No cephalomarginal setae.
Abdomen: Sclerotized areas distinct. Third tergites with (4-5) setae. Median hooks with 1 seta. Inner hooks with 2 apical setae, spine1/4, 2/3 total hook length. Fifth caudal tergites with

(25-28) setae; epipleura with (3-5) setae. Ninth eusternum with 2 groups of 3 setae on caudal margin; pygopod with 5 setae on each side.

Discussion

Overall, this species is most similar to *C. haemorrhagica* and *C. ocellata*, species with which it is taxonomically related. The first and second instars are nearly identical in most characters except for the small conical projection on the U-shaped ridge in *C. sedecipunctata* and *C. ocellata* and slightly lighter colors of some of the head appendages in *C. secedimpunctata*.

Notes

Current information on any citizen science project or organization mentioned in this book can be found easily on the Internet–just google the name or keywords. Much of my information, of course, came from personal communication. These endnotes mainly include sources for quoted written material or controversial facts, acknowledgement of original research or work, and some general reference books that I found particularly helpful. The e-mails in this text have been slightly modified for style, and the sender's address and any irrelevant information have been removed. If you have any questions about my sources or research, feel free to contact me at my website: www.sharmanaptrussell.com.

Introduction: Renaissance and Revolution

A history of Galaxy Zoo is on the website Zooniverse, including the numbers of early responses from citizen scientists. The site also gives information about current projects at Galaxy Zoo and lists the scientific papers their research has helped produce. The "apocryphal story" of their start can be found in many sources, but the one I relied on most was by Tim Adam, "Galaxy Zoo and the new dawn of citizen science," *The Observer*, March 18, 2012. In this engaging and informative account, Adam interviews Kevin Schawinski, the young astrophysicist in the bar, as well as his friend Chris Lintott, and identifies the bar itself as The Royal Oak on Woodstock Road in Oxford.

The idea that the universe might contain as many as 500 billion galaxies was estimated in 2009 by a German supercomputer simulation based on information from the Hubbell Space Telescope. Other estimates are more conservative, as in Erik A. Petigura, Geoffrey W. Marcy, and Andrew W. Howard, "Prevalence of Earth-size Planets Orbiting Sun-like Stars," *Proceedings of the National Academy of Science* 48, November 26, 2013, which used 200 billion galaxies to estimate the number of Earth-like planets in the observable universe–with 8.8 billion of those in the Milky Way alone. Yikes. The chances of life–us, plants, animals–being alone in the universe seem to be getting increasingly smaller. One among many or the only game in town: it's hard to say which scenario captivates the imagination more.

The best way to learn about citizen science programs like Foldit, National Geographic's Valley of the Khan Project, and the Cornell Lab of Ornithology's many projects is to google their names and start reading their websites. The website called SciStarter is also a clearinghouse for citizen science projects, large and small.

A good book that includes some history of citizen science is Janet Dickinson and Rick Bonney, editors, *Citizen Science: Public Participation in Environmental Research* (New York: Cornell University Press, 2012). I also learned more about this history at the Conference on Public Participation in Scientific Research, August 4–5, 2012 in Portland, Oregon and, in particular, from Abraham Miller-Rushing, Richard Primack, and Rick Bonney, "The history of public participation in ecological research," *Frontiers in Ecology and the Environment* 10 (2012), pages 285–290.

For the description of a Portland couple participating in Nature Notebook and Project Budburst, I am grateful to Lisa Gervier, a longtime member and enthusiast of these programs.

The high school students looking at fossils are participating in Fossil Finders, sponsored by the Museum of the Earth in Ithaca, New York. A discussion of pygmy hunters and gatherers using smartphones can be found at ExCiteS's website: www.ucl.ac.uk/excites. More about do-it-yourself-biologists can be found by googling DIYB and centrifuges–with blogs that start off tantalizingly with, "I had been thinking about building a centrifuge out of a blender but first I decided to research what other people had done before I tried my own hand at it."

July 2011

My main sources for information about tiger beetles were David L. Pearson, C. Barry Knisley, and Charles J. Kazilek, *A Field Guide to the Tiger Beetles of the United States and Canada* (New York: Oxford University Press, 2006) and David L. Pearson and Alfred P. Vogler, *Tiger Beetles: The Evolution, Ecology, and Diversity of the Cicindelids* (New York: Cornell University Press, 2001). I enjoyed and consulted John Acorn, *Tiger Beetles of Alberta: Killers on the Clay, Stalkers on the Sand* (Alberta: University of Alberta Press, 2001), which I bought mainly because I liked the title. There are other field guides to tiger beetles specific to locations around the world. I also subscribe to Ronald L. Huber, Ed., *Cicindela: a quarterly journal devoted to Cicindelidae*, available only by writing to Ron Huber, 2521 Jones Place West, Bloomington, MN 55431-2837. Ron is a well-known collector, expert, and citizen scientist in the field of tiger beetles.

The quote about being soothed and healed by nature is in the essay "The Gospel of Nature" by John Burroughs which can be found at the website http://www.readbookonline.net/readOnLine/21513/.

The quote concerning congenial work is from the essay "The Secret of Happiness," which can be found in John Burroughs, *The Writings of John Burroughs, Volume 12* (Boston: Houghton Mifflin, 1905), p. 279.

The Jane Goodall quote is from Jane Goodall, *Jane Goodall: Forty Years at Gombe* (New York: Steward, Tabon, and Chang, 1999), p.10.

The quote from Dick Vane-Wright is from Sharman Apt Russell, *An Obsession with Butterflies: Our Long Love Affair with a Singular Insect* (Boston, Mass.: Perseus Books, 2005), p.146.

A good source on mountain lions is Maurice Hornocker and Sharon Negri, Eds, *Cougar: Ecology and Conservation* (Chicago: University of Chicago Press, 2009). I also read a range of other books, with a range of attitudes, from Harley Shaw, *Soul Among Lions: The Cougar as Peaceful Adversary* (Tucson: University of Arizona Press, 2000) to David Baron, *The Beast in the Garden: the True Story of a Predator's Deadly Return to Suburban America* (New York: W. W. Norton Publishing Company, 2005).

The quote about the early Rio Grande is from Gaspar Pérez de Villagrá, *Historia de la Nueva México,* translated by Gilberto Espinosa (Los Angeles: The Quivira Society, 1933), which can be found online at http://nationalhumanitiescenter.org/pds/amerbegin/exploration/text1/villagra.pdf.

More information on the fish of the Rio Grande is in James Sublette, Michael D Hatch, and Mary Sublette, *The Fishes of New Mexico* (Albuquerque, New Mexico: University of New Mexico Press, 1990).

Like everyone else, I have rows of field guides on topics ranging from dragonflies to rocks, many of which I buy and put on the shelf hoping they will come to me—slipping down from their place, aloft and moving silently across the room—to infiltrate my dreams, hoping for osmosis and the power of proximity. One guidebook too heavy and substantial to do that is William G. Degenhardt, Charles W Painter, and Andrew H. Price, *Amphibians and Reptiles of New Mexico* (Albuquerque: University of New Mexico, 1996.) Any discussion of snakes or lizards has involved an exploration of that tome. On the other hand, one of the smallest field guides on my shelf is Pinau Merlin, *A Field Guide to Desert Holes* (Tucson, Arizona: Arizona Sonoran Desert Museum Press, 2003), which I have also consulted about the holes of New Mexican reptiles. I am sure, in fact, that this quirky little book does fly about my house at night.

The quote about the behavior of the Western red-bellied tiger beetle around water sources is from David L. Pearson, C. Barry Knisley, and Charles J. Kazilek, *A Field Guide to the Tiger Beetles of the United States and Canada* (New York: Oxford University Press, 2006), p.132. The quotes from the taxonomic key are from David L. Pearson, C. Barry Knisley, and Charles J. Kazilek, *A Field Guide to the Tiger Beetles of the United States and Canada* (New York: Oxford University Press, 2006), pp 24-36.

August 2011

The quote about snails is from an advance review copy of David Haskell, *The Forest Unseen: A Year's Watch in Nature* (New York: Viking, 2012), p.52 and the quote about the Mattole Indians is from David Abram, *Becoming Animal: an earthly cosmology* (Vintage, 2011), p. 92.

Information about plans to divert the Gila River is often local to southern New Mexico. As a board member of the Upper Gila Watershed Alliance (UGWA), partnering with other environmental groups like Gila Resources Information Project (GRIP) and Center for Biological Diversity, I have been engaged in this issue for many years. For detailed information, go to the Gila Conservation Coalition's website.

More of my own explorations of the Paleolithic lifestyle can be found in my novel *The Last Matriarch* (Albuquerque: University of New Mexico Press, 2000), set in southwestern New Mexico some eleven thousand years ago.

Information on rearing tiger beetles can be found in David L. Pearson, C Barry Knisley, and Charles J. Kazilek, *A Field Guide to the Tiger Beetles of the United States and Canada* (New York: Oxford University Press, 2006); David L. Pearson and Alfred P. Vogler, *Tiger Beetles: The Evolution,*

Ecology, and Diversity of the Cicindelids (New York: Cornell University Press, 2001); and John Acorn, *Tiger Beetles of Alberta: Killers on the Clay, Stalkers on the Sand* (Alberta: University of Alberta Press, 2001), as well as articles such as Rodger A Gwiazdowski, Sandra Gillespie, Richard Weddle and Joseph Elkinton, "Laboratory Rearing of Common and Endangered Species of North American Tiger Beetles (Coleoptera: Carabidae: Cicindelinae)," *Annuals of the Entomological Society of America* 104(3):534-542 (2011) and websites such as http://bcrc.bio.umass.edu/tigerbeetle. Ted MacRae's wonderful and thoughtful blog http://beetlesinthebush. wordpress.com also has information on rearing these beetles.

The quote about catching tiger beetles is from David L. Pearson and Alfred P. Vogler, *Tiger Beetles: The Evolution, Ecology, and Diversity of the Cicindelids* (New York: Cornell University Press, 2001), p. 254.

More information on Ed Greenberg's research on leaf concentrates can be found on the website http://www.leafforlife.org/.

September 2011

I have previously written about Carl Linnaeus and taxonomy in my nonfiction work *Anatomy of a Rose*: *Exploring the Secret Life of Flowers* (Boston: Perseus Books, 2003).

The arrow quote comes from *Jack Kerouac: Road Novels: 1957-1960* (Library of America, 2007), p. 25.

The information on larvae who cache their meals is from Harold L. Willis, "Bionomics and Zoogeography of Tiger Beetles of Saline Habitat in the Central United States (Coleoptera: Cicindelidae)," *The University of Kansas Bulletin* Vol XLVII, No. 5 (Oct. 11, 1967): 145-313.

Annie Dillard's quote is from Annie Dillard, *Pilgrim at Tinker Creek* (New York: Harper Perennial Modern Classics, 2007), p. 64.

October 2011

Coastal Wildscapes is a nonprofit group dedicated to preserving native plants and landscapes along the coast of Georgia. For more information, visit their website.

The quote on the physics of beauty is in Aldo Leopold, *Sand County Almanac* (New York: Ballantine Books, 1986), p.146.

More information on Terry Erwin and his early experiments can be found at the Smithsonian's National Museum of Natural History's website and in Chapter 12, *Rediscovering Biology*, http://www.learner.org/courses/biology/textbook/biodiv/biodiv_4.htm.

The material about Jeffrey Lockwood is from Annie Proulx, Ed., *Red Desert: History of a Place* (Austin: University of Texas Press, 2008).

For more information on Sky Island Alliance, go to their website and read, in particular, their ten-year report on programs and activities.

Much of my information about the middle-aged brain, including the studies cited, came from Barbara Strauch, *The Secret Life of the Grown-up Brain* (New York: Penguin Books, 2011).

Information about medieval trials of insects came from Ross H. Arnett Jr., Ed., *American Beetles,* Volume 2 (New York: CRC Press, 2002). More material about beetles being worn can be found at Victoria Z. Rivers, "Beetles in Textiles," e-zine *Cultural Entomology Digest* 1 http://www.insects.org/ced2/beetles_tex.html. More information about the religious uses of beetles, including Paleolithic beetles and beetles in Egypt, is available at Yves Cambefort, "Beetles as Religious Symbols" at e-zine *Cultural Entomology Digest* 2 at http://www.insects.org/ced1/beetles_rel_sym.html. Information about the natural history of some beetles came from many sources, including various field guides, textbooks, and personal communication.

The description of the third instar of the Western red-bellied tiger beetles is in Knisley, Barry C. and David L. Pearson, "The Biosystematics of Larval Tiger Beetles of the Sulfur Springs Valley, Arizona: Descriptions of Species and a Review of Larval Characteristics for *Cicindela (Coleoptera: Cicindelidae,"* Transactions of the American Entomological Society Vol.110 (Dec. 31, 1984), p. 539.

Winter 2012

I first wrote about diapause in butterflies in *An Obsession with Butterflies: Our Long Love Affair with a Singular Insect* (Boston, Mass.: Perseus Books, 2005).

The quote concerning Foldit is by Justin Siegel, a post-doctoral student working with one of the founders of Foldit, from Jessica Marshall, "Victory for crowd-sourced biomolecule design," *Nature News* (January 22, 2012) at http://www.nature.com/news/victory-for-crowdsourced-biomolecule-design-1.9872. The quote on the successes of the Mastodon Matrix Project is from Robert M. Ross, et al, "The Hyde Park Mastodon Matrix Project with Particular Reference to Mollusks and Seeds," *Palaeontographica Americana* 61 (2008), p.126.

For more information on climate change in the Southwest, I recommend William deBuys, *A Great Aridness: Climate Change and the Future of the American Southwest* (New York: Oxford University Press, 2011).

The quote in the Xerces publication is David Pearson, "Six-legged Tigers," *Wings* 34.1 (2011), p. 23.

The quote about a woman who sits and sees is from Lawrence Stapleton, Ed. *H.D. Thoreau, A Writer's Journal*, (Dover Publications, 2011), p. xii. The quote about waking up is from Annie Dillard, *Teaching a Stone to Talk: Expeditions and Encounters* (New York: Harper-Collins, 1988), p. 22.

Spring 2012

Two guidebooks to plants I could not do without are Jack C. Carter, Martha A. Carter, and Donna Steven, *Common Southwestern Native Plants* (Silver City, New Mexico: Mimbres Publishing, 2009) and Jack C. Carter, *Trees and Shrubs of New Mexico* (Silver City, New Mexico: Mimbres Publishing, 2012).

My sources on mountain lions include Maurice Hornocker and Sharon Negri, Eds, *Cougar: Ecology and Conservation* (Chicago: University of Chicago Press, 2009); Harley Shaw, *Soul Among Lions: The Cougar as Peaceful Adversary* (Tucson: University of Arizona Press, 2000); and David Baron, *The Beast in the Garden: the True Story of a Predator's Deadly Return to Suburban America* (New York: W. W. Norton Publishing Company, 2005).

The quote by Lucille Tower is from "A Big Day for Science: Citizens Have Contributed One Million Observations to Top Nature Database," *USGS News*, May 3, 2012 at http://www.usgs.gov/newsroom/article.asp?ID=3195#.U1p3rFf9WRI.

June 2012

News about the Whitewater-Baldy Fire was local and ongoing.

The quote about tiger beetle collectors not being very sophisticated enemies is from David L. Pearson and Alfred P. Vogler, *Tiger Beetles: The Evolution, Ecology, and Diversity of the Cicindelids* (New York: Cornell University Press, 2001), p. 167.

For material about coatis, I would like to thank Christine Hass, with whom I personally corresponded. Hass is the author of numerous past articles on coatis and a future book. The quotes about coatis are taken from Bil Gilbert, *Chulo: A year among the coatimundis* (Tucson: University of Arizona Press, 1973), pp. 185, 142-43, 113.

Information about tiger beetles comes mainly from David L. Pearson, C. Barry Knisley, and Charles J. Kazilek, *A Field Guide to the Tiger Beetles of the United States and Canada* (New York: Oxford University Press, 2006) and David L. Pearson and Alfred P. Vogler, *Tiger Beetles: The Evolution, Ecology, and Diversity of the Cicindelids* (New York: Cornell University Press, 2001).

As with many things, I first learned about benign and malignant envy on Wikipedia. Then I went to their sources, including Niels Van de Van, et al., "Leveling up and down: the experiences of benign and malicious envy," *Emotion* 3, June 9, 2009: 419-429. Apparently, people in the Netherlands already have two words for these two different states. The study also included populations in Spain and the United States.

The quote from the essay "The Gospel of Nature" by John Burroughs can be found in *The Writings of John Burroughs: Time and Change* (Boston: Houghton, Mifflin, and Company, 1912), pp 251-252, a Google e-book.

July 2012

One of my favorite books on clouds is Gavin Pretor-Pinney, *The Cloudspotter's Guide* (New York: Berkeley Publishing Group, 2006).

The idea about the universe reflecting on itself is from Thomas Berry, *The Dream of the Earth* (Sierra Club Books, 2006).

I have not written a great deal about my father except for the title essay in *Songs of the Fluteplayer* (Addison-Wesley, 1991; reprinted by University of Nebraska Press, 2001); the essay "Letter to my Father Concerning the State of the World," *Terrain.org* 23 (Winter/Spring 2009); and the flash nonfiction "Icarus," *The Baltimore Review* (Fall 2013).

August 2012

The quote about the punctured tiger beetle is from David L. Pearson, C. Barry Knisley, and Charles J. Kazilek, *A Field Guide to the Tiger Beetles of the United States and Canada* (New York: Oxford University Press, 2006), p.122.

An interesting book on the amount of data now available on the internet is David Weinberger, *Too Big to Know* (New York: Basic Books, 2012).

To augment my tracking workshop, I used James C. Halfpenny, *Scats and Tracks of the Desert Southwest* (Helena, Montana: GlobePequot Press, 2000) and Paul Rezendes, *Tracking and the Art of Seeing* (New York: Harper Collins, 1999).I would also like to thank Janice Przbyl, the leader of the workshop, and program coordinator for Wildlife Linkages from 2001-2009.

The quote by fire ecologist Stephen J. Pyne, as well as other material about the Wallow Fire, is from Ron Dungan, "After the devastation, nature fuels recovery, season by season," *The Republic,* June 22, 2012 at http://archive.azcentral.com/travel/articles/2012/06/04/20120604wallow-fire-arizona-one-year-later.html.

The quote about wolves from "Thinking Like a Mountain" is in Aldo Leopold, *Sand County Almanac* (New York: Ballantine Books, 1986), p.138.

A good article on the immediate effects of the Whitewater-Baldy Fire is Neil LaRubbio, "What scientists are learning from wildfire in New Mexico," *High Country News*, Nov. 12, 2012.

The quote and material about Patty Jo Watson can be found in my nonfiction book *When the Land was Young: Reflections on American Archaeology* (Boston, MA: Addison-Wesley Press, 1994; reprinted University of Nebraska Press, 2001), pp 73-74.

September 2012

I first wrote about memory in butterflies in *An Obsession with Butterflies: Our Long Love Affair with a Singular Insect* (Boston, Mass.: Perseus Books, 2005).

Information on weeds and invasive species came from many sources. I've long been interested in the medicinal uses of plants and first wrote about that, as well as about phytoremediation, in *Anatomy of a Rose: Exploring the Secret Life of Flowers* (Boston: Perseus Books, 2003).

Information on sunflowers at Fukushima can be found at Molly Cotter, "Millions of Sunflowers Soak Up Nuclear Radiation in Fukushima" at http://inhabitat.com/thousands-of-sunflowers-soak-up-nuclear-radiation-in-fukushima/ as well as other newspaper and internet articles.

For Southwestern archaeology, I reread a number of good books, including J.J. Brody, *Mimbres Painted Pottery* (Albuquerque: University of New Mexico Press, 1977) as well as reading the newer Linda S. Cordell and Maxine E. McBrinn, *Archaeology of the Southwest* (Walnut Creek, California: Left Coast Press, 2012) and Stephen Lekson, *A History of the Ancient Southwest* (Santa Fe: SAR, 2009). I also drew on experiences and research from my *When the Land was Young: Reflections on American Archaeology* (Boston, MA: Addison-Wesley Press, 1994; reprinted University of Nebraska Press, 2001).

The quote about burrowing owls is from Pinau Merlin, *A Field Guide to Desert Holes* (Tucson, Arizona: Arizona Sonoran Desert Museum Press, 2003), p.118. Another good book is the unique and fascinating Charley Eiseman and Noah Charney, *Tracks and Signs of Insects and Other Invertebrates* (Mechanicsburg, PA; Stackpole Books, 2010).

October 2012

More about the Site Steward program in New Mexico can be found on their website http://sitestewardfoundation.org/.

Index

Note: Italic page numbers refer to photographs.

scratches on sycamore tree, 187–189, *188*

Sea Floor Explorer, 184–185

Shaw, Harley, *A Soul Among Lions: Cougar as Peaceful Adversary*, 106

side-oats grama, *65*

Site Steward program (USFS), 186, 190–193

Sky Island Alliance
  Burro Mountains tracking project, 157, 169
  tracking teams, 158
  Wildlife Linkages, 75–76, 154–156

smartphone apps, 15, 90, 151

SnowTweets Project, 88

Sonoran mountain king snake, 124

A Soul Among Lions: Cougar as Peaceful Adversary (Shaw), 106

specimen preservation, 76

splendid tiger beetle, 72

St. Catherine's Island (Georgia), 69–70

Stardust@Home project (NASA), 9, 54

stink bug (*Halyomorpha halys*), 59

T

Tanzania, community-based conservation projects in, 93

tarantula, 177

Tasmania, tiger beetles in, 112–113

taxonomic keys, 34–35

taxonomy, history of, 60

"Thinking Like a Mountain" (Leopold), 156

thin-lined tiger beetle (*Cicindela tenuisignata*), 60

Thoreau, Henry David, 98, 147–148

tiger beetles
  as bio-indicators of biological diversity, 15, 72
  characteristics of, 25–26
  Cicindela genus, 87–88
  fishing for larvae, 32–33
  instar phase, 64
  larvae, stages of, 17–18, 31–32

locations of, 17

National Museum of Natural History collection, 71–72

number of species, 17

oviposition, 47

predators of, 118–120, 177

in salt lake habitats, 29

senses of, 125

stilting, *126*

survival underwater, 165

techniques for catching, 23–24

tunnels of, 32

worldwide collections of, 72

See also specific species

Tiger Beetles of the Western Hemisphere (Erwin and Pearson), 71

Tiger Beetles: The Evolution, Ecology, and Diversity of the Cicindelids (Pearson and Vogler), 14–15, 118

tortoise beetle, 79

Tower, Lucille, 110

Truth or Consequences, New Mexico, 61

U

Uganda, community-based conservation projects in, 93

University of Massachusetts, website on beetle mating, 47

Upper Gila Watershed Alliance, 106

US Army Corps of Engineers, 132

US Forest Service, 158
  Mexican spotted owl study, 121–124
  Site Steward Handbook: Find It, Record It, Report It, 192
  Site Steward program, 186, 190–193
  use of Birds in Forested Landscapes, 92

US Geological Society, 150
  See also Nature's Notebook

US National Weather Service, 147

USA National Phenology Network
  IceWatch USA, 88, 98
  Nature's Notebook, 55–56, 92